MAN-APE APE-MAN

The Quest for Human's Place in Nature
and
Dubois' 'Missing Link'

First published in The Netherlands in 1993 by the Netherlands Foundation for Kenya Wildlife Service,
Pieter de la Court Building, Wassenaarseweg 52, 2333 AK Leiden, The Netherlands

This book was designed and produced by Maseland Grafische Vormgeving,
Tollenstraat 30, 4101 BE Culemborg, The Netherlands

ISBN: 90 263 1285 7

Printed by Chevalier Holland Printers, P.O. Box 210, 3340 AE Hendrik Ido Ambacht, The Netherlands

Distribution for The Netherlands: Uitgeverij Ambo BV, Baarn
Distribution for Belgium: Uitgeverij Westland NV, Schoten

CENTENNIAL PITHECANTHROPUS

NNM

stichting VSB Fonds Den Haag en Omstreken

MAN-APE APE-MAN

The Quest for Human's Place in Nature
and
Dubois' 'Missing Link'

Richard E. Leakey L.Jan Slikkerveer

Foreword by H. R. H. Prince Bernhard of The Netherlands

All proceeds of this book are destined for:

KENYA
WILDLIFE
SERVICE

NETHERLANDS FOUNDATION
FOR KENYA WILDLIFE SERVICE

CONTENTS

Foreword by H. R. H. Prince Bernhard of The Netherlands	7
Acknowledgements	9
Introduction	11
1 The Discovery of Pithecanthropus Erectus	23
2 Evolution and the Emergence of Ecology	35
3 Dubois' Passion for the 'Missing Link'	53
4 *Lemuria*: Journey to The Netherlands East Indies	73
5 Disclosure and Debate: Man-Ape or Ape-Man	91
6 Africa: The 'Cradle of Mankind'	113
7 *De Bedelaer*: New Perspectives on 'Man 's Place in Nature'	137
8 The Past is the Key to Our Future	165
References	179
Illustration Credits	180
Index	182

FOREWORD

B

Soestdijk Palace, June 1993

Our understanding of human's present evolutionary and ecological position on planet Earth has advanced substantially over the past decades, and a growing concern over the rapid habitat destruction and its implications for complex forms of life has gained a strong scientific basis. As a result, habitat loss is now the confirmed predominant factor threatening wildlife survival, and as such, ultimately the survival of humankind.

Advanced scientific research and modern technology, however, have significant historical roots that go back much deeper in time to the days of late nineteenth-century pioneers such as Charles Darwin, Thomas Henry Huxley and Ernst Haeckel.
While evolutionary theory had already emphasized the crucial role of the interaction between living organisms and their environment, indicating that species would become extinct when ecological conditions changed too rapidly, a total reconsideration in western philosophical thinking was introduced with the concept that humans were to be regarded as a part of nature.

It was the Dutchman Eugène Dubois who, in his scholarly pursuit of the advance of knowledge of man's place in nature started a century ago to look to the past, to the fossil record of the former Netherlands East Indies, to find the 'Missing Link' between us and the rest of nature. Although his famous discovery of the fossil remains of the 'Ape-Man from Java' made him one of the 'Founding Fathers' of palaeoanthropology, he was also in fact among the first 'ecologists' in Holland to see that the establishment of a wildlife sanctuary in his *'Natuur Park De Bedelaer'* in South Limburg would set the tone for the nature conservation movement in The Netherlands for the century yet to come.

It is encouraging that today, one century after Dubois' disclosure of his discovery, the reassessment of his pioneering work and broadening vision from the past - as Richard Leakey and Jan Slikkerveer document so cogently and tellingly in this book - has induced questions about human origins and development to be posed within their ecological context. "In what way have human populations been interacting with wildlife communities, and how can the protection of habitats and their biodiversity contribute to the changes required for our future adaptations to planet Earth?"

Today, we still face the same problems of survival that our ancestors faced, albeit in a different dimension: our ultimate dependence - despite our knowledge and technology - on the animal and plant kingdoms on this Earth. Clearly, however, Dubois' early vision of our place in nature has indicated the direction in which a solution can be found: the achievement of a new sustainable balance with the natural world in which the conservation of endangered species and their habitats will be the most critical issue in environmental conservation efforts in the years to come.

Bernhard

Prince of the Netherlands

ACKNOWLEDGEMENTS

"In 1993, won't it be a hundred years ago since Eugène Dubois broke the news to the world, that he had found the first 'Missing Link' between Man and Ape?" "Yes, it is, and we should do something about it".

The place was Muthaiga, a residential area in Nairobi, and the time was December 1988. The conversation between H.E. Raden Kusumasmoro (the Indonesian Ambassador to Kenya at the time) and both authors of this book did in fact lay a significant basis for the Pithecanthropus Centennial 1893-1993. As a culmination of several commemorative meetings and conferences in Indonesia, France, and Germany, the Centennial would eventually focus on the wider significance of Dubois' work and vision on 'Man's Place in Nature', encapsulated in the general theme of "Human Evolution in its Ecological Context".

Since that time, numerous individuals and institutions have contributed to the various activities including an Exhibition *MAN-APE APE-MAN, Pithecanthropus in het Pesthuys,* an International Congress encompassing four interdisciplinary Sessions on Palaeoanthropology, Cultural Anthropology, Human Ecology, and Philosophy, and an Excursion to Dubois' 'Nature Park' *De Bedelaer* in the south of Holland. This book, that seeks to mark the centennary of Dubois'pioneering work in this context, could not have been realized without the advice and contribution of many colleagues and friends. We would like to thank Dr. Mohamed Isahakia, Dr. Geoffrey W. Clarfield, Ms. Christine Kabuye and Mr. Nelson Tabu of the National Museums of Kenya in Nairobi for their advice and research materials. Also, we would like to express our gratitude to Ms. Lynette Anyonge of Kenya Wildlife Service in Nairobi for her very kind assistance at times of pressure.

We are equally in the debt of Prof. Phillip V. Tobias of the University of Witwatersrand, Johannesburg, South Africa, Prof. John R.F. Bower of the University of Minnesota in Duluth, USA, and Dr. Alan Barnard of the University of Edinburgh, UK for their academic advice.

We wish to express our gratitude for the use of the unique research materials on Dubois' scholarly work to Bert Theunissen and Ton Lemaire and for the provision of the crucial information on the *Collectie Dubois* in Leiden by the present Curator, Dr. John de Vos.

We are grateful to Dr. Kees P.C.M. Oomen, Chairman of the Board of Leiden University for his generous support of this publication. Equally, we would like to express our gratitude to Mr. Jacques van Vliet for his continuous advice for the finalization of the book. From Leiden University, we also thank Dr. Wim H.J.C. Dechering, Drs. Henk Korenromp, Mr. Dick Kaphorst, Drs. Wouter Teller, Mr. Willem Arentshorst, and Mr. Loek A. Zuyderduin for their copious support. We highly appreciate the contributions of Prof. Pieter Baas and Mr. W.J. Holverda of the *Rijks Herbarium* in Leiden. We are grateful to Dr. Henk W.J.G. Schalke, Dr. Hans Beijer, and Dr. Raymond Corbey, members of the Pithecanthropus Centennial Foundation, as well as to Dr. Gerard J. van den Broek, Dr. Dick A. Hooijer, and Drs. Marvin E. Egberts for their valuable inputs. From the National Museum of Natural History in Leiden, we also acknowledge the substantial contribution of Drs. Wim van der Weiden, Mr. Alex Scholten, Dr. Mary Bouquet, Mrs. Isabelle Galy, and Prof. L.B. Holthuys.

Without the endeavour of several descendants of the Dubois Family in Holland such as Mrs. Anneke Hooijer, Mr. Jean M.F. Dubois, Ir. Louk E.M. Dubois, Dr. Emile and Mrs. Albertine Dubois, and Mrs. Victorine Heukensfeldt Jansen-Dubois, this book would never have reached the personal dimension of the many-sided figure of Eugène Dubois.

This publication would never have been realized in its present form without the generous financial support of V.S.B. Bank Den Haag en Omstreken and the unceasing efforts of Mr. Evert Chevalier and his Printing Office, Mrs. Ingrid Maseland and her Associates, and Mrs. Rosemary Robson-McKillop B.A. (Hons.).

Finally, we would especially like to express our gratitude to Drs. Mady Slikkerveer for her personal efforts and advice.

01 Eugène Dubois, 1858-1940.

INTRODUCTION

One century ago, in 1893, Eugène Dubois, a young anthropologist/anatomist from The Netherlands, officially announced his new world-famous discovery of the first 'Missing Link' between Man and Ape. The discovery he made on the banks of the Solo River in East Java, Indonesia, included a fossil skullcap, thigh-bone, and molar - presently kept in the vault of the National Museum of Natural History in Leiden, The Netherlands. His disclosure shocked both the scientific world and the general public of the late nineteenth century.

If Charles Darwin's *The Origin of Species* (1859), Thomas Huxley's *Man's Place in Nature* (1863), and Ernst Haeckel's *Natürliche Schöpfungsgeschichte* (1868) had already earlier shaken Victorian society to its foundations by expounding the new, evolutionary theories on human evolution, Dubois became the first to provide unique fossil evidence of our age-long and direct relationship with the natural world. Although the publication of his study of the fossil remains of an early ancestor, to which he attributed the scientific name *Pithecanthropus erectus* or the 'Upright-walking Ape-Man', marked the beginning of a new stage in the reconstruction of the history of humankind, the true significance of Dubois' work and vision has so far remained rather outside the limelight: his evidence of a new perspective on our place *within* - and not *above* - nature.

02 The famous Pithecanthropus fossils shown for the first time in 100 years to the general public during the 1993 exhibition MAN-APE APE-MAN. Pithecanthropus in 'Het Pesthuys' in Leiden.

Many palaeoanthropologists have come to regard Dubois as one of the 'founding fathers' of their field, mainly

03 *Postage stamp of 1967 from Cuba, depicting Pithecanthropus erectus. Portrait is based on the late Czech artist Zdeněk Burian's reconstruction.*

because of his discovery of the first fossils of what now is called *Homo erectus*, and his related work in cephalization research. His dedication of the greater part of his life to the holistic study and practice of human evolution in its ecological context and his scholarly pursuit of, in his own last written words (1940:1275): "*...the advance of knowledge of man's place in nature, what is commonly called human phylogenetic evolution...*" has, in fact, made him a pioneer in what has only lately emerged as the most critical field of the science of ecology: the interaction between the human species and its environment.

Indeed, his fascination from an early age with the flora, fauna, and the composition of the soil, and his eagerness to learn the Latin names of all the trees, shrubs, grasses, and mosses on the banks of the River Maas near his birthplace of Eysden in the south of Holland already held the promise of his future scholarly work in this, at the time newly-developing, field of study of our place in nature from prehistoric times up to the present.

Greatly inspired by the evolutionary theories of Darwin, Wallace, Huxley, and Haeckel during his student years, Dubois who was a genuine outdoor type of person decided to undertake fieldwork in the tropics instead of pursuing a successful career as an anatomist at the University of Amsterdam. Encouraged by Haeckel's idea that early human ancestors had originated from *Lemuria*, a sunken continent thought to be near India, he declined a professorship and left the university to go to one of the Dutch colonies: The Netherlands East Indies, now Indonesia, albeit against the advice of many of his friends and colleagues.

On 29 October, 1887, accompanied by his wife Anna and their infant daughter, he embarked on what has become the greatest scientific exploration in the history of human evolution and ecology: a journey by steamer to The Netherlands East Indies. After his first expeditions into the hinterland of Sumatra, in 1890 he went to Java where, in the period between 1891 and 1892, he crowned his extensive fossil collection of numerous plants and animals with the discovery of the now famous remains of the Ape-Man from Java.

Then, in 1893, Dubois decided to make an official announcement of the discovery of *Pithecanthropus*, claiming it to be the first tangible 'Missing Link' between Man and Ape.

Back in The Netherlands in 1895, Dubois' publications, lectures, and pre-

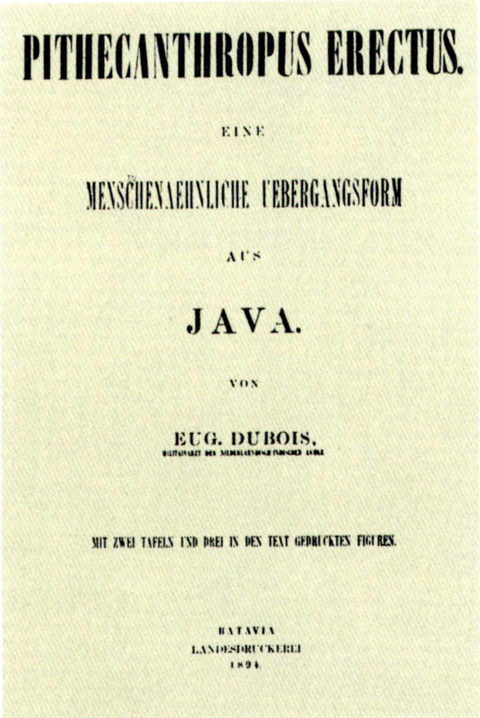

PITHECANTHROPUS ERECTUS.

EINE

MENSCHENAEHNLICHE UEBERGANGSFORM

AUS

JAVA.

VON

EUG. DUBOIS,

MIT ZWEI TAFELN UND DREI IN DEN TEXT GEDRUCKTEN FIGUREN.

BATAVIA
LANDESDRUCKEREI
1894.

04 *Frontispiece of Dubois' provocative publication in 1894 on his discovery of the 'Missing Link', Pithecanthropus erectus, the Ape-Man from Java.*

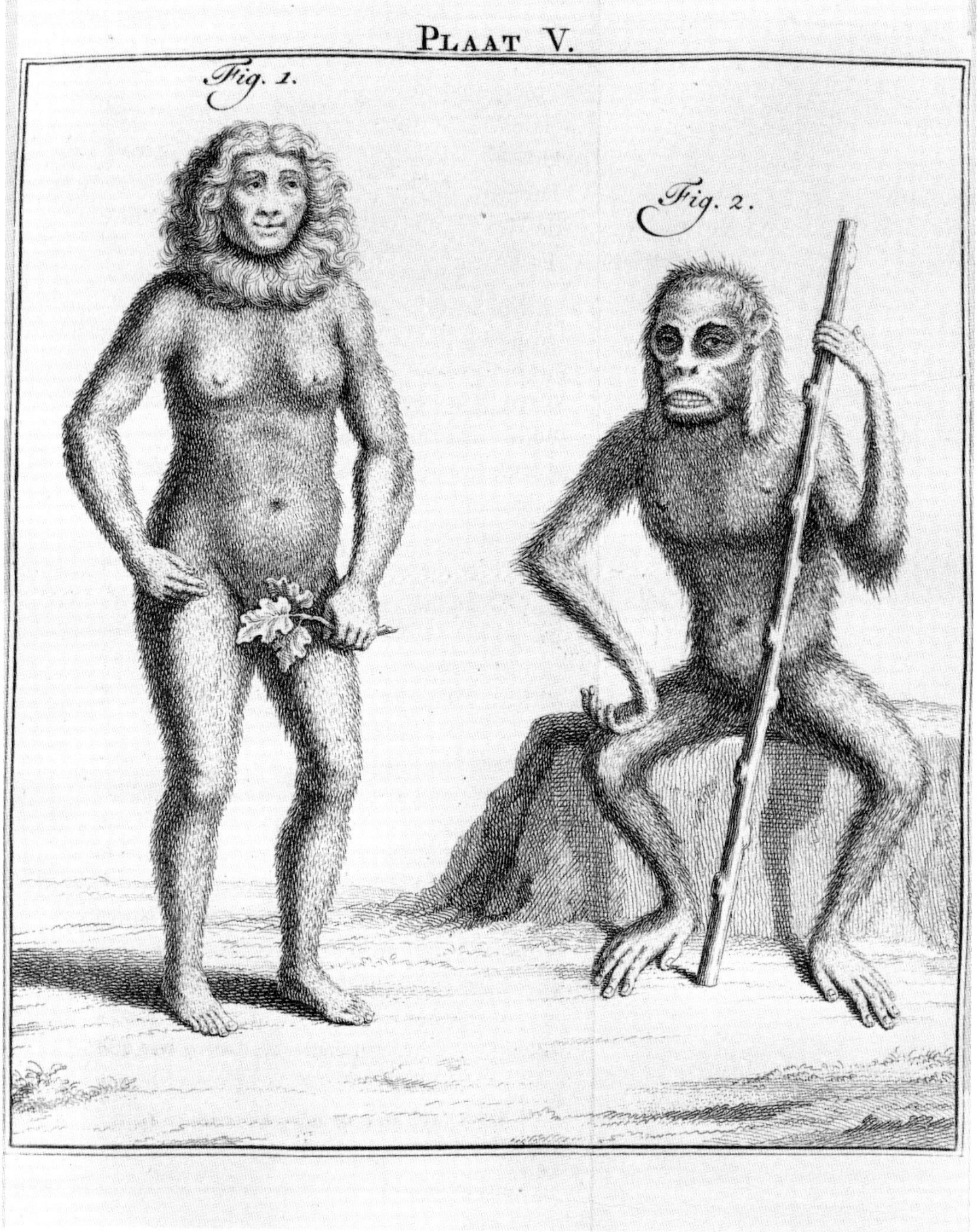

05 *Early images of primates used to exaggerate the human-like characteristics, as is the case with these redrawings by later artists of the Orang-utan of Bontius (1658, on the left), and Tyson (1699, on the right).*

sentations at several meetings in Leiden, Liège, Brussels, Paris, London, Edinburgh, Dublin, Berlin, and Jena caused heated debates and stirred up a real controversy about the humanoid fossils from Java and their place in the human 'family-tree'. Were his fossils to be attributed to an Ape, a Human, or to a creature in between: either a Man-Ape or an Ape-Man?

Scepticism about Dubois' interpretation of the fossils prevailed among the numerous reactions from leading scientists of his time, such as Krause (1895), Virchow (1895), Lydekker (1895), Martin (1895), Cunningham (1894-1995), Keith (1895), and Turner (1895); while only a few tended to endorse his views, including Manouvrier (1895), Jaekel (1895), Marsh (1896), and Haeckel (1894-1896). Only later on, in the course of the early 20th century, was Dubois' claim to gain more general credibility, prompting a new generation of anthropologists to study human origins in the Far East and, later, elsewhere around the globe. The same sort of debate would eventually precede all the famous discoveries that still lay ahead in Europe, in South Africa, and notably in East Africa.

Meanwhile, around 1900, embittered by the mounting opposition to his views, for nearly a quarter of a century Dubois locked the Java fossils in a vault, away from collegial doubters, jealousies, and negative responses. He withdrew to work on his collection of fossil vertebrates and on cephalization, but above all to put Haeckel and Darwin's ideas on ecology into practice. While the church reproached him from the pulpit and the public joked about 'our descent from the apes', his retreat into rural Limburg fuelled wild legends to explain his flight into oblivion.

06 Pithecanthropus erectus has fascinated the general public since its discovery. This part of the Mexican comic 'Eugenio Dubois' of 1959 highlights its mystery in the Museo Nacional.

In fact, Dubois decided that until his own new description of the Pithecanthropus fossils appeared, no other scientists could have further access to them. It was only in the 1920s, that the *Koninklijke Nederlandsche Academie voor Wetenschappen* (Royal Netherlands Academy of Arts and Sciences) finally persuaded him to make a concession. Eventually, the summary of his study on the ape-like characteristics of his fossils, published in 1935 under the title *On the Gibbon-like Appearance of Pithecanthropus erectus*, created confusion as many of his contemporaries asserted that Dubois had suddenly revised his standpoint, redescribing Pithecanthropus as an Ape. The reverse was true, as Theunissen (1989) has elucidated, Dubois argued yet

07 The typical moor landscape of South Limburg in Holland, where Dubois executed his later, pioneering studies in ecology.

again, that Java Man stood halfway between Ape and Man, and as such at the same distance from both.

Gradually Dubois settled into a sedentary life concentrating on the impassioned study and practice of the newly-developing field of ecology on his estate *De Bedelaer* near Haelen in Limburg. Using the comparative study of fossil seeds and woods he retrieved from the Tegelen clay-layers from his estate as a basis, he imported numerous seeds of exotic trees and shrubs from different parts of the tropical world, of the type which had grown in Limburg hundreds of thousands of years before. These included the unique mammoth tree and the swamp cypress, still flourishing there today.

After changing the water level and the lake of *De Bedelaer*, the re-introduced 'prehistoric' vegetation returned this unique Limburg landscape of more than 40 hectares partly to its primordial form. His experiments fascinated many famous naturalists and conservationists including Jac P. Thysse and Eli Heimans, who visited Dubois regularly.

Complementing his fossil evidence approach and his 'law of cephalization' as entirely new means to substantiate the theory of human's direct relation to nature, his practical fieldwork finally took shape. With a true naturalist's zeal, he recreated *De Bedelaer* into a prehistoric landscape in Limburg that served to make our past even more tangible and relevant to our understanding of the evolution process of *Homo sapiens* in relation with the environment. Meanwhile, he also became one of the pioneers in The Netherlands movement for the conservation of nature. When the new concept of '*Natuurmonument*' ('Monument of Nature') was introduced to keep and conserve certain parts of Holland as a 'monument' to the past, Dubois sought to designate his estate accordingly.

On 16 December, 1940, shortly before his 83rd birthday, he died on his estate, that as a monument to his work has continued to attract the attention of many ecologists and conservationists up to the present day. His famous collection of numerous unique books and manuscripts, faunal, floral, and human fossil remains, as well as as an assemblage of brain casts of a variety of animals was transferred to the National Museum of Natural History in Leiden, where as a part of the *Collectie Dubois* it has become a key element in the ongoing pursuit of the study of Man and Nature.

This book is an account of his historic journey of discovery, his work on *De Bedelaer* and their far reaching implications for the exploration of new ideas that go beyond the search for human origins: our understanding of human's place in the natural world, and its direct relevance to our present ecological problems.

It comes out at a time - one century after Dubois' provocative publication on *Pithecanthropus* in 1893 - in which much work on our distant origins has been and still is carried out in Asia and, notably during the past fifty years, in Africa. Spectacular discoveries and new directions in palaeoanthropology, based on evidence from ecology and molecular biology, have extended our knowledge of the early evolutionary history of our ancestors, particularly with regard to the *Australopithecus, Homo habilis,* and *Homo erectus* in their African context. Indeed, Dubois' Homo erectus has now been established as an early ancestor the first to use fire and the first to leave Africa as the 'Cradle of Mankind'.

08 *Centennial medal with the logo of the "Pithecanthropus Centennial 1893-1993" and the portrait of Eugène Dubois with a beard during his stay in Indonesia. Designed and executed by the Dutch sculptor Frank Letterie, Vorden.*

Nonetheless, our understanding of what has lately become a prominent topic, the origin of modern human, *Homo sapiens,* and its place in the universe has only very recently come to the fore.

Even more so - a century after the official establishment of the term 'ecology' by the International Botanical Congress in 1893 - this account will re-introduce the holistic perspective of Dubois on our relationship with the environment, with the human milieu, which since his day has come to witness a profound reassessment of what then had been called the 'philosophy of living nature'.

Dubois' primary contribution has certainly been his approach to the reconstruction of human evolution on the basis of hominid fossil remains. Although we now know that Dubois' *Pithecanthopus* belongs to the more human-like species of *Homo erectus* and, as such, is not a true intermediate form between Man and Ape, undoubtedly, his 'Java Man' introduced a new scientific image of Man as a mere 'Evolutionized Ape'.

A reassessment of his philosophy as a naturalist and his later work in conservation and ecology - enhanced by Ton Lemaire's revealing studies (1977-1978) on *De Bedelaer* - sheds new light on the ultimate objective of his life. Pithecanthropus as such was not an aim in itself, rather it was only a means to an end. His true contribution to the scientific endeavour of his time touched upon a much wider concern: to substantiate man's closer evolutionary relationship with the animal kingdom. In Thomas Henry Huxley's words "*Man's Place in Nature*". In sum, his great merit has certainly been that he introduced a new perspective to provide the first tangible proof of the great antiquity of humankind: our 'Missing Link' with Nature.

In view of such great achievement, the centenary celebration of Dubois' pioneering work has focused the 'Pithecanthropus Centennial 1893-1993' on the appropriate theme of

09 *Painted view of Homo erectus after a reconstruction of the human fossils from Solo by the late Czech artist Zdeněk Burian.*

Human Evolution in its Ecological Context. Including an International Conference at Leiden University, an Exhibition *Man-Ape Ape-Man; Pithecanthropus in Het Pesthuys* in the National Museum of Natural History in Leiden, an Excursion to his Natural Park *De Bedelaer* in Limburg, and several scientific and cultural activities in collaboration with institutes such as the National Museums of Kenya and Kenya Wildlife Service in Nairobi (Kenya), and Universitas Padjadjaran, Bandung (Indonesia), the Centennial has sought to encompass a reflection of the past contributing towards the future of our species.

This commemorative book begins with a personal account of the actual discovery of the 'Missing Link' in the banks of the Solo River near Trinil in the island of Java, as it was recorded by Dubois' own son, Jean M.F. Dubois. This is followed by a sketch of the changing scientific decorum of eighteenth century evolutionist theory and the emergence of the concept of '*oecologie*' in the framework of which Eugène Dubois lived and worked. A description of his youth, education, and growing passion for the 'Missing Link' and his determination to find it in The Netherlands East Indies is the prelude to an account of his laborious but rewarding expeditions into the hinterland of Sumatra and Java.

Following an overview of the heated debate he created with the disclosure of the fossils in 1893, it continues by examining the implications of Dubois' pioneering vision for the steps that lay ahead on the road of reconstruction, of human evolutionary history. Towards the end, new perspectives stemming from Dubois early interpretation of the place of our species are presented, based on new evidence, approaches, and ideas. In Louis Leakey's historical words: "*The past is the key to our future*".

As a tribute to Eugène Dubois and to all those who have followed in his footsteps in the study of human evolution and ecology in different parts of the world, we hope that by reassessing his pioneering work and vision from the past our perspective on 'Human's Place in Nature' may be extended into the future for the ultimate survival of humankind in the universe.

010 *Logo of the Pithecanthropus Centennial Foundation 1893-1993 in Leiden, The Netherlands.*

011 *Centennial gentleman's ties in red and navy blue, emblazoned with Dubois' own original nineteenth-century reconstruction of the Pithecanthropus erectus skull and a 'life' reconstruction of the Ape-Man from Java. Designed by Richard Leakey and Mady Slikkerveer.*

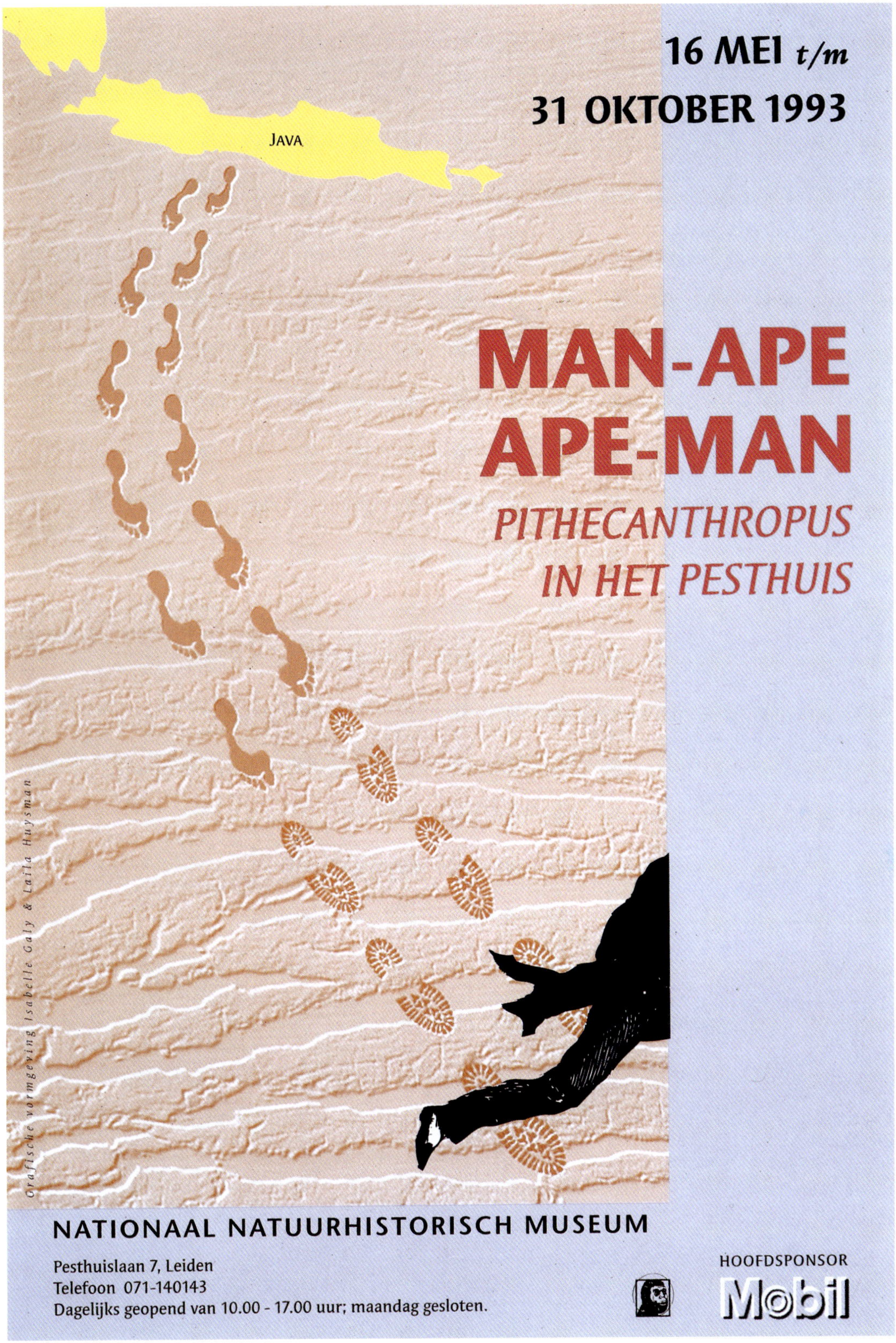

012 Poster of the Exhibition MAN-APE APE-MAN. Pithecanthropus in het Pesthuys, 15 May - 31 October 1993, Leiden. Designed by Isabelle Galy and Laila Huysman.

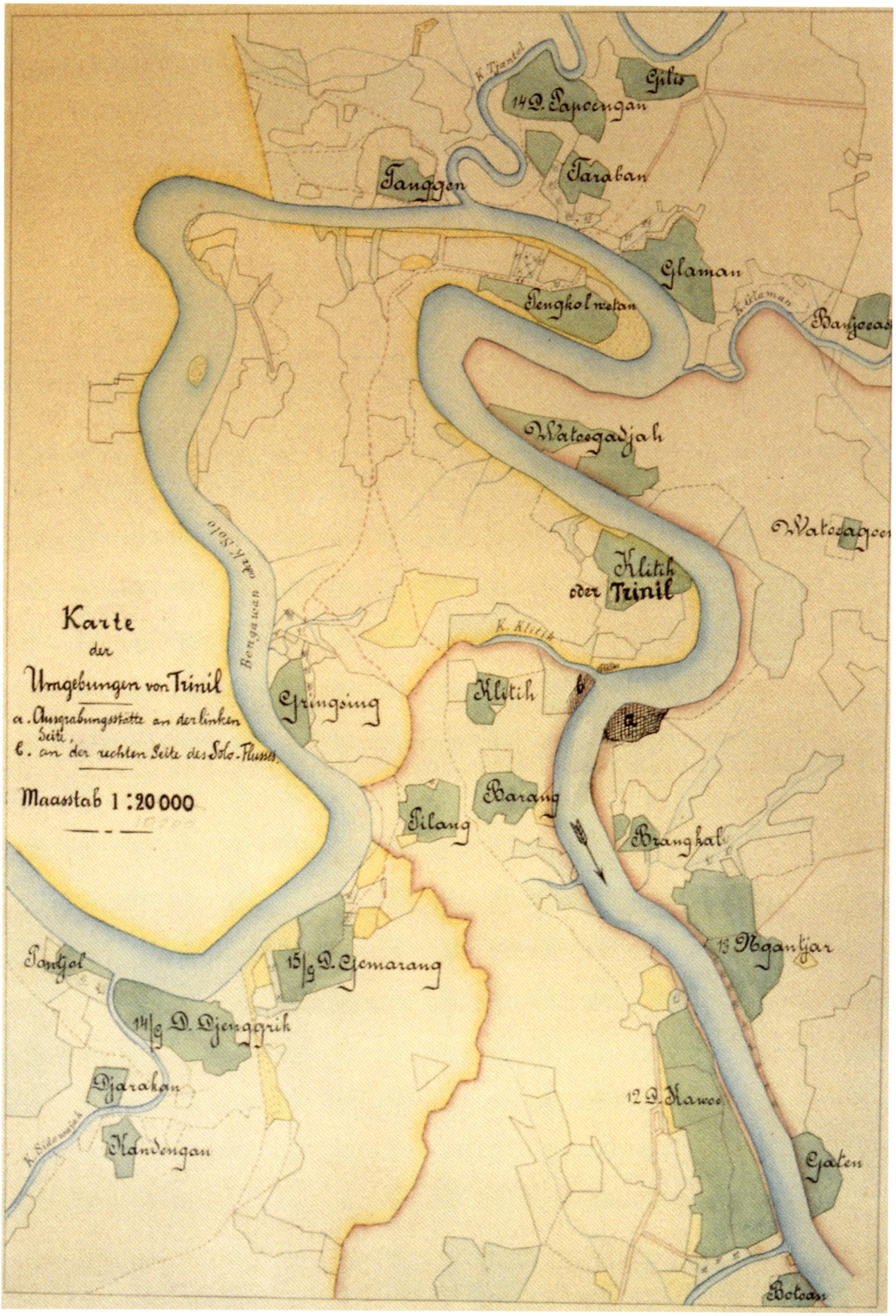

1.1 *Map of Trinil and surroundings in East Java, made by Dubois.*

1 The Discovery of Pithecanthropus Erectus

High in the blue tropical sky a flight of wild ducks passed over the river. Wongsosemito looked up at them and, leaning on his hoe, remarked "Soon we can go home. The rains will come and the river will rise." Sardi, who came from the same village and was working next to him, grinned. "*Baiklah,* That's good", he commented. With a wider grin which displayed an array of betel-blackened teeth, he added, "Yes, the water will cover all this, and there will be no more work."

"*Baiklah*", said Wongsosemito as his hoe, *patjol*, dug into the soft sandstone of the river bank. A few more times the hoe rose and fell.

Then it struck a rock.

There had been many of them, but this time Wongsosemito looked more closely at the dark object near his feet. He had unearthed many stones with his *patjol* before, but somehow this one seemed different. Sardi thought so too. Leaning heavily on their hoes, they discussed the matter in great detail, for this was a weighty matter. Finally, the two squatted down to have a closer look at the stone. Wongsosemito touched it with his fingers, then slowly brushed some of the soil from it. The rock revealed a rather smooth surface, this was a new cause for discussion. This rock *was* different, and they had better bring it to the attention of their *mandur* (foreman).

So they called Ahmad Saleh, the *mandur,* and the three men squatted down around the strange-looking rock and discussed it some more.

The year was 1891; the scene, a river-bank in Java. A group of Javanese labourers were digging into the hard soil of the bank, just above the level of the stream of the *Bengawan Solo* or Solo River, which was at its usual low for this time of the year, the dry season. Nearby two Dutchmen were watching them. They were the Sergeants Kriele and De Winter of the Engineering Corps of the *Koninklijk Nederlandsch-Indisch Leger, KNIL* (Netherlands East Indies Army), who were in charge of the excavation. Under the supervision of the two Dutchmen, the forced labourers toiled silently, their brown bodies glistening in the heat of the tropical sun, their *patjols* rising and falling methodically onto the sedimentary rock with dull thuds.

1.2 Local fishermen on the Solo River near Trinil around 1900.

1.3 Sergeants Kriele and De Winter of the 'Koninklijk Nederlandsch-Indisch Leger' (KNIL), who supervised the excavations at Trinil.

1.4 Main gate of the former Dutch Military Fortress Van den Bosch near Ngawi on the Solo River in East Java, from where a century ago the excavations in Trinil were organized. Unfortunately, the fortress has recently fallen into decay (November 1991).

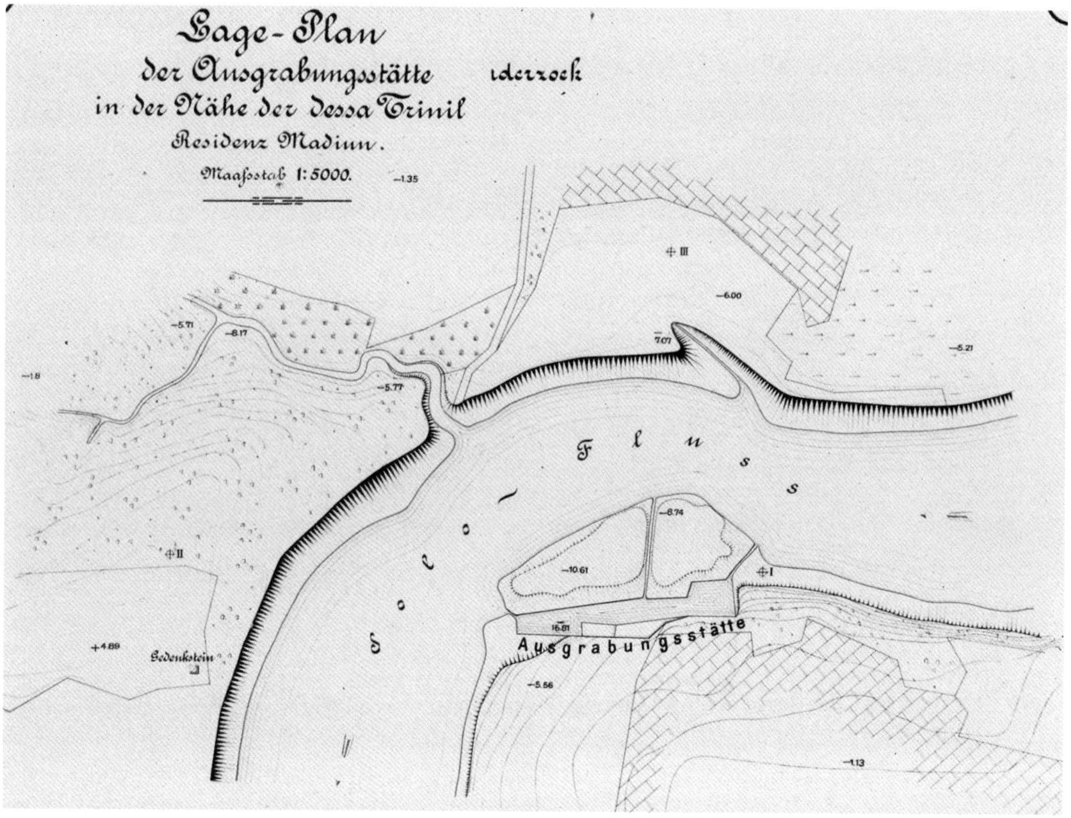

1.5 Detailed map of Dubois' excavation site on the banks of the Solo River near Trinil in East Java.

1.6 *Eugène Dubois and his wife Anna Lojenga at the time of their stay in East Java.*

The excavations were undertaken from the nearby *Fort Van den Bosch,* a military fortress at Ngawi on the Solo River that was part of the Dutch defensive works along the *Grote Postweg* (Great Mailroad) connecting West with East Java.

Occasionally one of the men would momentarily drop his hoe and pick up a stone or fossilized bone. Reverently, as was the custom of the Javanese people at the time, he would hand his find to one of the sergeants, then return to his work, quietly, unhurriedly.

A little further down the bank, at the edge of the excavation, was another Dutchman. He was young, in his early thirties, and of medium stature though powerfully built. His handsome face, with steel-blue eyes, firm mouth, and short moustache and beard, bore the ruddy complexion of the outdoor man. While the sergeants showed a tendency to unbend and lean against the steep side of the bank in a most unmilitary manner, the other man's bearing was that of an officer inspecting his troops as he walked from one toiling labourer to another. There was no doubt who was in command.

1.7 A group of forced labourers engaged in the excavations near Trinil.

As he approached both sergeants he asked: "Did they find anything unusual?" The words were soft-spoken, but firm and to the point. Sergeant Kriele straightened up, then shook his head. "No, nothing new, Doctor", he replied, and pointing to a small pile beside him, added, "The same old stuff, Sir, mostly stones and a few bones." The doctor made his way to the pile and examined a few of the fossil bones. "Same old stuff," Kriele repeated.

At that moment Ahmad Saleh came running towards them. He was not a member of the prison gang to which the labourers belonged, but came from the neighbouring village, Ngawi. Active and intelligent, he made a good *mandur* and he could speak Malay, the language the Dutchmen understood.

Ahmad Saleh squatted down in the local fashion and said, "*Saya minta ampun, Tuan.* (I beg your pardon, Sir)". "*Mau apa?* (What do you want)," Sergeant Kriele asked sharply.

After the usual pauses, which the custom demanded from one addressing a superior, the Javanese told him that one of his men had unearthed a strange looking rock."*Itulah sama kura-kura, adanya.* (It is like a tortoise and that is the truth)," Ahmad Saled said.

"Where is it?" asked Kriele. "*Di situ, Tuan* (Over there, Sir)," the *mandur* replied, rising slowly, his body slightly bent and, walking backwards, he respectfully led the way.

The Dutchman saw at once that it was unlike any of the other fossils they had found in the sandstone formation of the river bank during the many weeks of excavation work. The "rock" was like a medium sized land tortoise as Ahmad Saleh had said.

The young doctor looked at it for some time and made notes of the location where it had been found, taking measurements, and collecting samples of the surrounding soil. Then,

1.8 The banks of the Solo River during the excavations by Dubois between 1891-1893.

with the help of Kriele, he cautiously removed the fossil from its age-old bedding. There was no doubt that it was a fossil, quite heavy and almost chocolate-brown in colour. He slowly turned it in his hands, then very carefully scratched off some of the impacted stone matter that encrusted the fossil, and he wondered about the unusual shell of the tortoise.

Many fossil remains of reptiles, mostly crocodiles, had been found at the excavation, but this was the first fossil tortoise unearthed. The inside of its "shell" was entirely filled with stone matrix, which accounted for the excessive weight of the fossil.

For a long time the young man studied it, and the more he looked, the more it puzzled him. If it was not a tortoise, what was it? It might be anything, the knee cap of a prehistoric elephant or part of the skeleton of some other as yet unknown animal. Still, it might be a smooth-carapaced tortoise, and yet, it did not look like one.

Perhaps initially somewhat confused, upon closer examination Dubois was soon pervaded by the feeling that this fossil was something entirely different. The rather heavy, dark brown object he was holding in his hands, just as it had been unearthed from the river-bank would soon become the most widely debated if not most controversial discovery in the early study of human evolutionary history.

Since its first official disclosure in a publication in 1893 as *Pithecanthropus erectus*, the Ape-Man from Java, Dubois' long-sought 'Missing Link' between Man and Ape, has fuelled both fact and fiction which have persisted up to the present time.

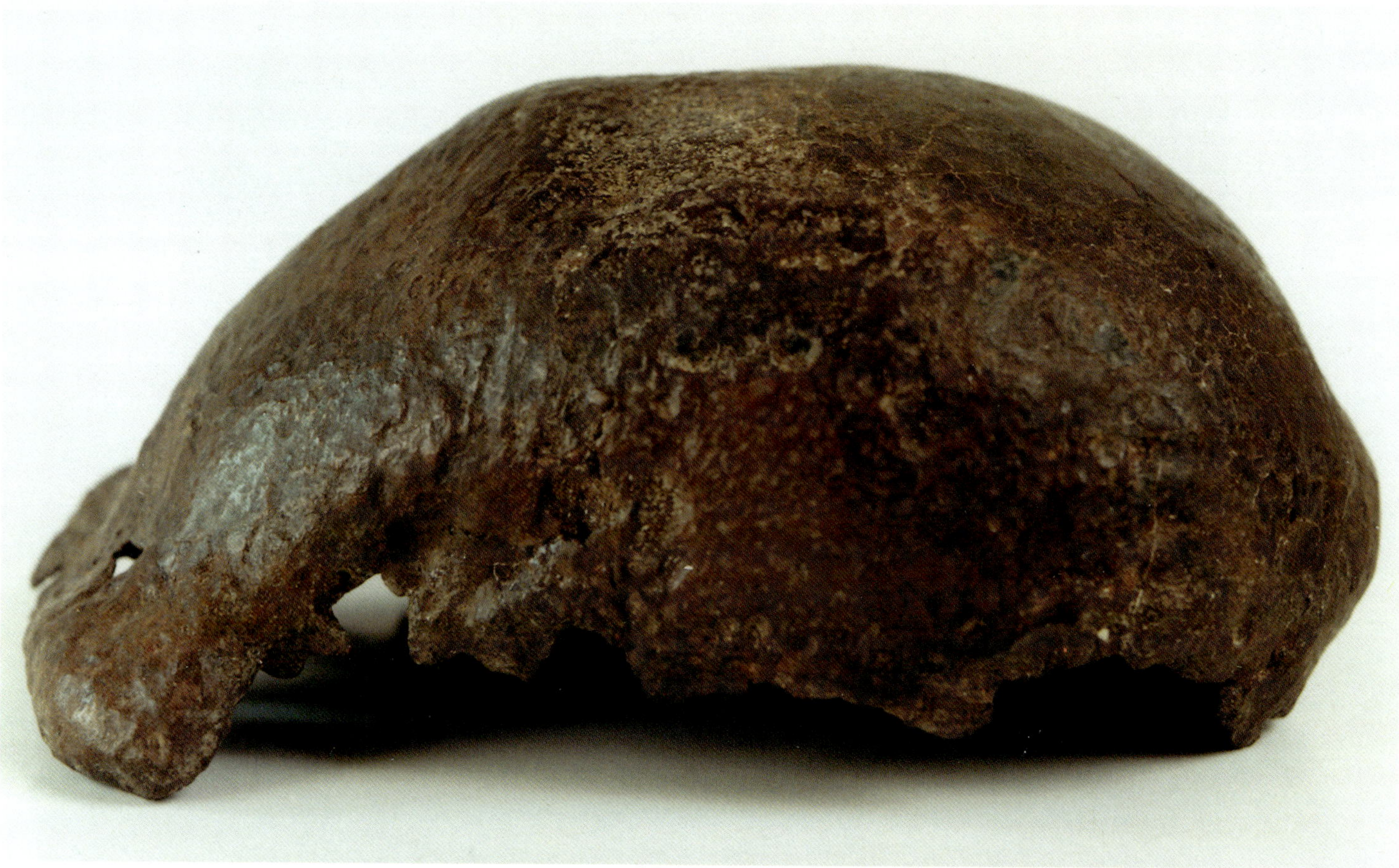

1.9 *'Life-size' photograph of the famous skullcap of Pithecanthropus erectus, unearthed near Trinil in Java in October 1891.*

1.10 *The left thigh-bone of Pithecanthropus erectus, unearthed near Trinil in August 1892.*

1.11 The present state of the find-spot in the banks of the Solo River near Trinil (November 1991).

1.12 The present state of the River Solo near Trinil (November 1991).

1.13 The local population of Trinil has always taken a keen interest in the find-spot of the Pithecanthropus erectus. (November 1991).

2.1 Charles Darwin in 1840. Watercolour by George Richmond.

2 EVOLUTION AND THE EMERGENCE OF ECOLOGY

Dubois was born and raised at a time in which conventional views that the Creator had designed each species with Man at the top were being challenged by Victorian science implying that the Human Race was part of the Animal Kingdom.

Charles Darwin's revolutionary book *On the Origin of Species by Natural Selection*, first published in 1859, and many times reprinted, annotated, and translated into more than thirty languages, has raised havoc among both scientists and the general public ever since. Although the author was not the first among the nineteenth-century scientists to put forward the notion that animal and plant species have evolved and changed over millions of years, his name is directly connected with the development of science as are those of Copernicus, Newton, or Aristotle.

Darwin's major contribution has not only been his remarkable scientific approach to the substantiation of evolutionary theory, but also the explanation he offered for the complex process of change of species: the concept of *natural selection.* As a young man of twenty-two, in 1831 he had boarded H.M.S. *Beagle* travelling around the world as a naturalist to collect and describe the fauna, flora, and geological formations of South America, South Africa, and Australia, and the islands of the Pacific and South Atlantic.

During this long voyage, after retrieving many marine fossils at an altitude of 12,000 feet in the Andes and huge bones of the arroyos of the Pampas, he visited the unique Galapagos Islands. In this the unspoiled 'evolutionary laboratory' of our planet with its giant tortoises and many different species of finches, whose beaks varied according to their food, Darwin found strong evidence of natural selection in combination with a changing environment. Back home in his study at Down House in Kent, he elaborated on the ideas of his predecessors about organic evolution, including those of his own grandfather, Erasmus Darwin. Combining the observations made during his voyage on *The Beagle* with such earlier accounts on biological evolution, including *An Essay on the Principle of Population* by Thomas Malthus (1797) and the *Phylosophie Zoologique* by the French naturalist Jean Baptiste Lamarck (1809), Darwin finally provided the theory with the explanatory basis of the concept of natural selection. Although he first formulated this important idea in 1838 - without officially publishing it then - by coincidence another naturalist, the Welsh botanist Alfred Russsel Wallace, arrived at the same notion independently in 1858.

Wallace, who had been collecting rare tropical plants and animals in South America and

2.2 Nineteenth-century cartoon of Charles Darwin offering a mirror to an ape.

South-East Asia for private collectors and museums, sent Darwin a short account of his ideas that year entitled *On the Tendency of Varieties to Depart Indefinitively from the Original Type,* without realizing that Darwin had already discovered natural selection. As a result of their correspondence and through the mediation of their colleagues Charles Lyell and Joseph Hooker, their joint paper was presented in a most generous way to the Linnaean Society as a *Darwin-Wallace Memoir on Natural Selection* (1859).

As the process of evolution seemed to proceed extremely slowly, the theory was well supported by Lyell's *Principles of Geology* (1830-1833) that - in contrast to earlier estimates that the age of our planet was not older than 6,000 years old e.g. as suggested by Bishop Ussher - provided evolution with an adequate time dimension.

Darwin's work unequivocally postulates that the world of animals and plants has evolved through a long and gradual process of variation and natural selection, and does not result from a once-only act of creation. With regard to the explanation of human origins, he was careful to mention only at the end of his book *Origin of Species*, that: *"Light will be thrown on the origin of man and his history."* However, the implications of evolutionary theory for the position of Man were obvious: Man belongs to the same world as the animals, and has developed from 'lower' forms through the same process of variation and natural selection. The question of human ancestry quickly led to the general conclusion that our distant ancestors must be sought among the larger primates, and other, non-scientific considerations of religion and ethics soon came to dominate the debate, ridiculing Darwin and our 'descent from the apes' in many satires and jokes.

Meanwhile, about three years after the publication of Darwin's momentous book *Origin of Species*, his friend Thomas Henry Huxley published *Evidences as to Man's Place in Nature* (1893-1894), in which he provided new proof that there were basically no anatomical differences between humans and apes. Based on research in comparative anatomy between humans and apes and in embryology he postulated a close evolutionary relationship between humans and the great apes. As the American author Roger Lewin (1989:1) recently quite

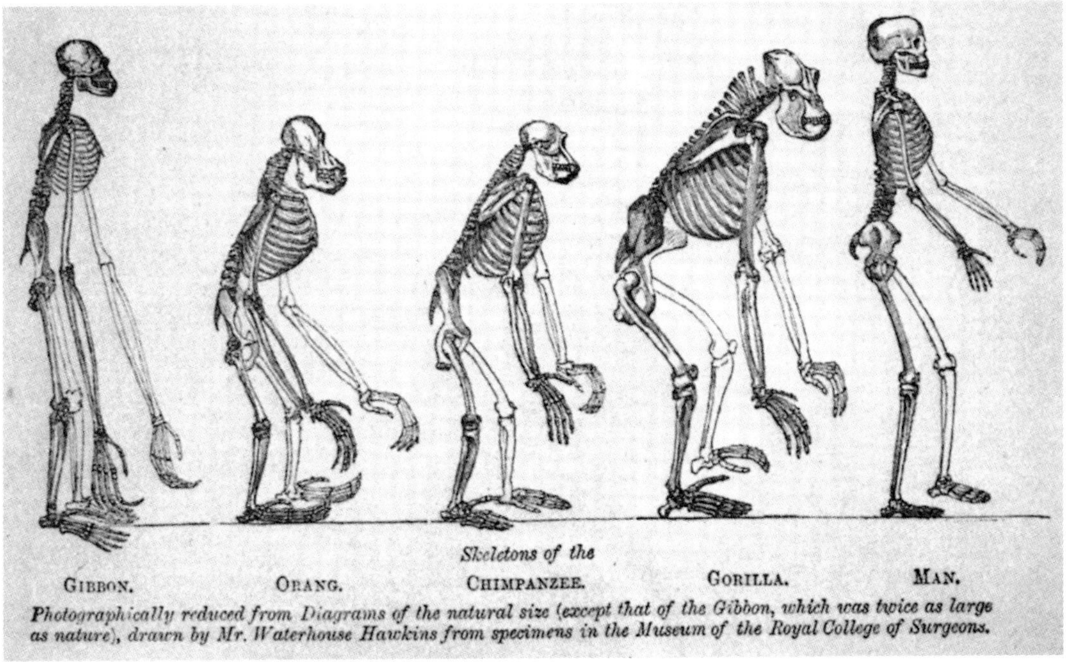

2.3 "Whatever part of the animal fabric might be selected for comparison, the lower Apes (monkeys) and the Gorilla would differ more than the Gorilla and Man." Illustration from T.H. Huxley's 'Man's Place in Nature' (1863).

rightly noted, Huxley's most significant conclusion on the close relationship with humans : *"...was a key element in a major revolution in the history of western philosophy: humans were to be seen as being* ***a part of*** *nature, no longer as* ***apart from*** *nature."*

The famous debate between Thomas Henry Huxley and Richard Owen on the presupposed superiority of human intelligence over apes as a result of certain characteristics of the human brain was clearly decided in favour of the former. However, an overriding issue in the discussion of the position of Man remained the human mind, which was still regarded as a feature separating humans from animals. As a result, the focus of attention in the debate on the 'gap' between animals and plants soon shifted to the 'gap' between humans and apes. In their efforts to bridge this gap, many scientists sought to over-expose either the humanness of the apes or the pithecoidness of some 'lower' human races. Several narratives were produced about upright-walking, dressed-up apes, using humans for slaves, and even mating with them for offspring. More accounts from explorers, colonizers, missionaries, and adventurers were given about their 'contacts' with ape-like humans - mostly in the 'dark continent of Africa' - who behaved as savages or barbarians without the benefit of morality or culture.

Indeed, as a result of both the growing perception of the unity of the natural world on the one hand, and the awareness of human variation based on these early accounts of 'strange' peoples overseas on the other, the history of humans descending from the primeval pair in Paradise became hard to maintain. Historians and natural scientists started to develop classifications describing these 'strange' peoples as partially developed, ape-like *Homo sapiens.* In the 'great chain of being' so-called Hottentots and Bushmen were easily placed in a sub-human category. This was based on the European explorers' perceptions of their appearance

a

2.5 The comparative anatomical study of apes and monkeys continued to play a key role in the development of evolutionary theory, in which the ecological aspects soon were incorporated. This illustration of G.H. Schubert (1882) includes the Barbary Ape, the Long-tailed Monkey, the Baboon, the Mandrill and the Howler Monkey.

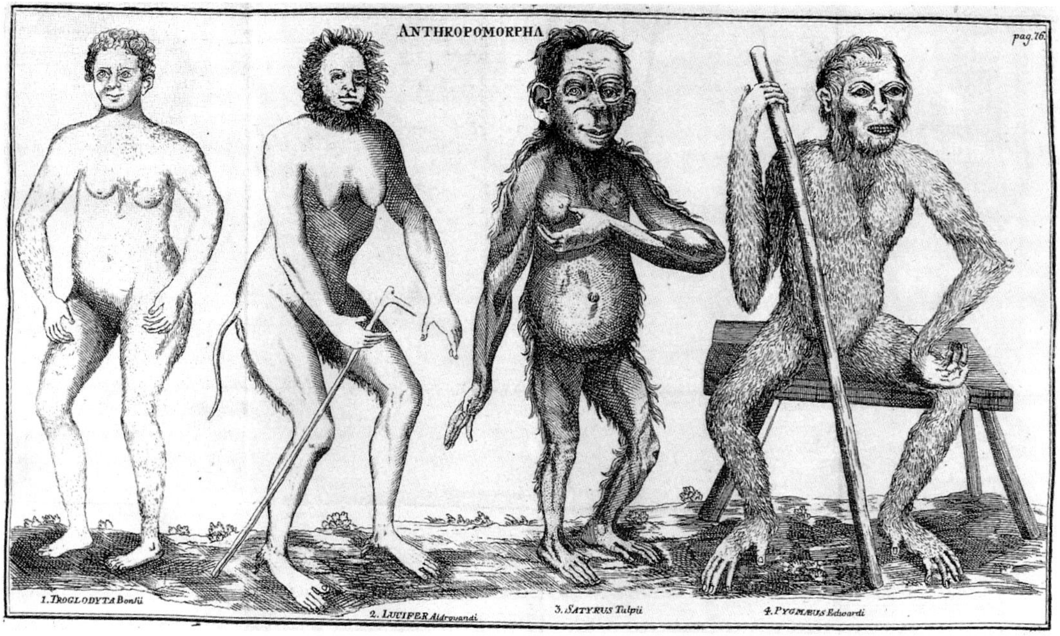

2.4 Four supposed 'manlike' apes classified by Linnaeus (1735). From left to right there are two imaginative creatures, the Chimpanzee (Satyrus tulpii), and the Orang-utan (Pygmaeus edwardi).

2.6 The German philosopher and anatomist Ernst Haeckel (1834-1919).

Page 41
2.7 Part of a fold-out plate of W. Buckland (1836) showing a diagrammatic section of the earth's crust. It shows the ecological relation between early rock formations and their characteristic fossils, reconstructed in drawings as 'living' plants and animals.

SECONDARY SERIES
Land Plants.
Mammifers Reptiles and Insects.
Marine Animals and Plants
New Red Conglomerate
Magnesian Limestone
Shell Limestone
Variegated Sandstone
Variegated Marl
Muschelkalk
Lias
Oolite
Green Sand
Carboniferous Limestone
Old red Sandstone
Transition Conglomerate
Clay Slate

and language as being similar to that of 'monkeys who: "chattered in the rainforest".

Early classifications of humans as part of the primate order had originated from the mid-eighteenth century formal classification system of Carolus Linnaeus. In his major publication *Systema Naturae* (1735), in addition to *Homo sapiens* this famous Swedish botanist included unknown, 'lower' creatures such as *Homo troglodytes* and *Homo caudatus*. The assumption that certain basic elements of humankind were of great antiquity gradually transpired from additional classificatory schemes constructed by Linnaeus' followers such as Buffon (1749), Blumenbach (1781), and Cuvier (1790).

Meanwhile, in a subsequent publication *The Descent of Man*, Darwin (1871) also expatiated on human evolution from ape-like ancestors encompassing a process of both physical and mental evolution. Here, he elaborates on the position of Man within the animal kingdom, leading to the conclusion that humans have evolved from more ape-like ancestors through a gradual process of evolution. At that time the African apes, the chimpanzee and the gorilla, were regarded as Man's closest relatives, while the Asian apes, such as the orang-utan were placed in a separate group.

The study of living primates would later gain more attention from scientists in search of our ancestors, particularly with regard to their behavioural pattern, their social organization, and the ecological niches in which they lived.

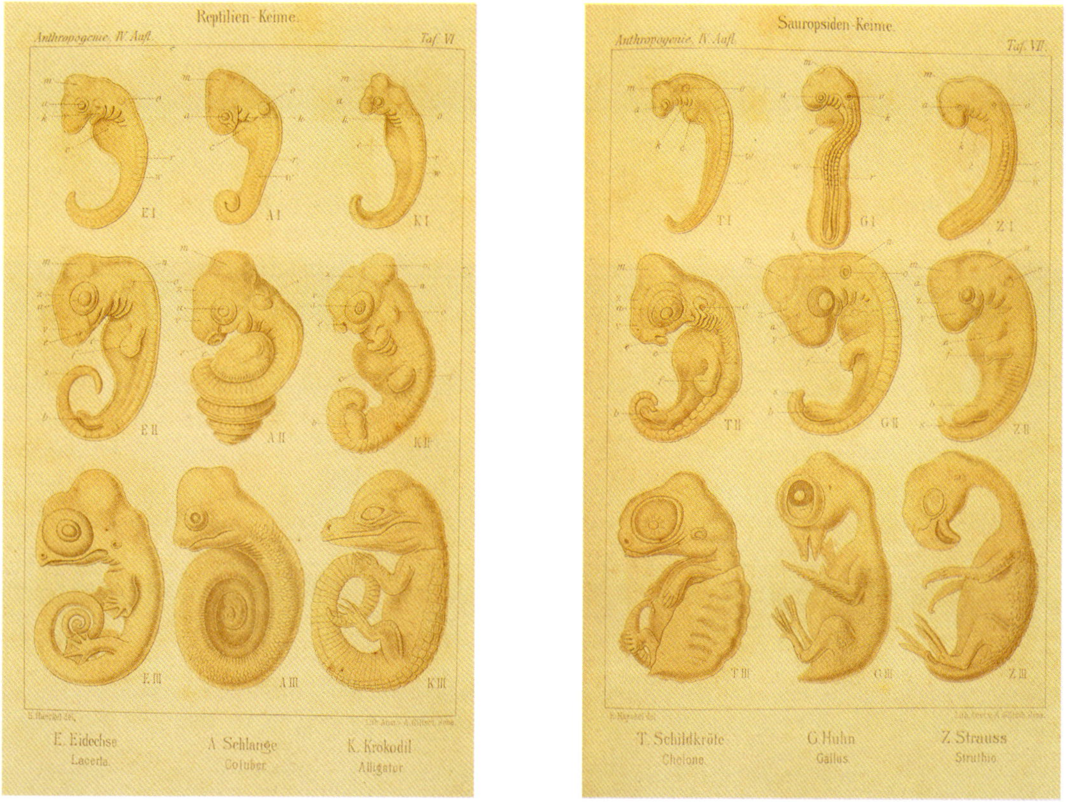

2.8 Embryos of lizzard, snake and crocodile (left) and tortoise, chicken and ostrich (right), at three stages in their development, as illustrated by Ernst Haeckel (1894-1896).

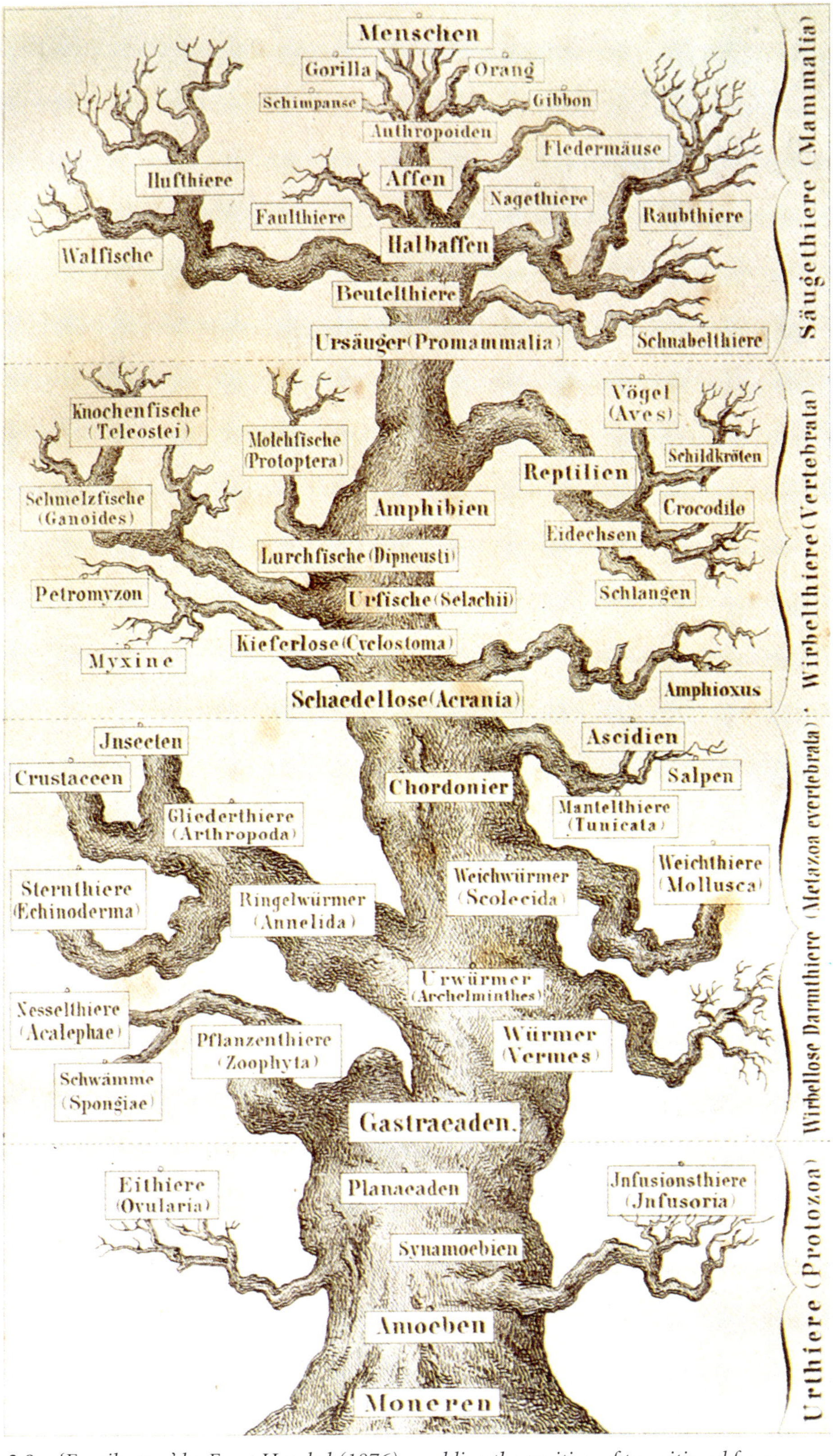

2.9 'Family-tree' by Ernst Haeckel (1876), enabling the position of transitional forms between the Apes (Affen) and Humans (Menschen).

2.10 The evolutionary chart on the 'Modern Theory of the Descent of Man' from Ernst Haeckel's 'Antropogenie oder Entwicklungsgeschichte des Menschen' (1876).

Supported by Huxley, whose statement "The structural differences which separates Man from the Gorilla and the Chimpanzee are not so great as those which separate the Gorilla from the lower Apes" soon became known as 'Huxley's Law'(1893-1894). By the end of the century the concept of human evolution gradually gained ground both in and outside Great Britain. As it was Darwin who first and foremost had brought so many different kinds of information - heredity and variation, fossils, geological formations, geographical distribution, embryology, taxonomy and homology - to bear on the question of evolution soon found many influential champions on the continent, during his lifetime and still today.

In Germany, Darwin's theories were integrated into the 'developmental philosophy' of Ernst Haeckel, who clearly observed the analogy between general processes of evolution and the development of the embryo to maturity. In his *Natürliche Schöpfungsgeschichte: Gemeinverständliche Wissenschaftliche Vorträge uber die Entwicklungslehre im Allgemeinen und Diejenige von Darwin, Goethe und Lamarck im Besonderen,* Haeckel (1868) developed a kind of nineteenth-century holism, often referred to as 'monism', in which he argued that matter and spirit were manifestations of the same underlying substance. In this way, monism fostered a general respect for life as a whole, substantiating the 'Unity of Nature'. Here, it is important to note that such a holistic, evolutionist perspective has laid the foundation for a vivid awareness of environmental interdependence. According to a recent study by the historian Peter J. Bowler on *The Environmental Sciences* (1992), such a vision: "...would have gravitated towards the environmentalist movement." The interest in the direct relation between the physical environment - geology and mineralogy - and the living species both in prehisto-

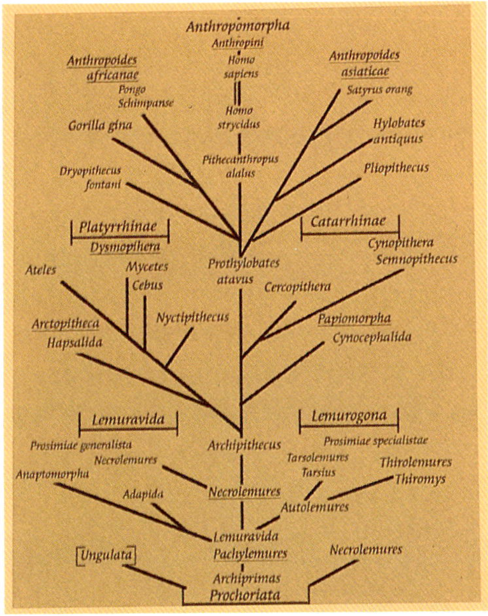

2.11 The human phylogeny of Ernst Haeckel (1874), already indicating the intermediate position of the by then not yet found Pithecanthropus.

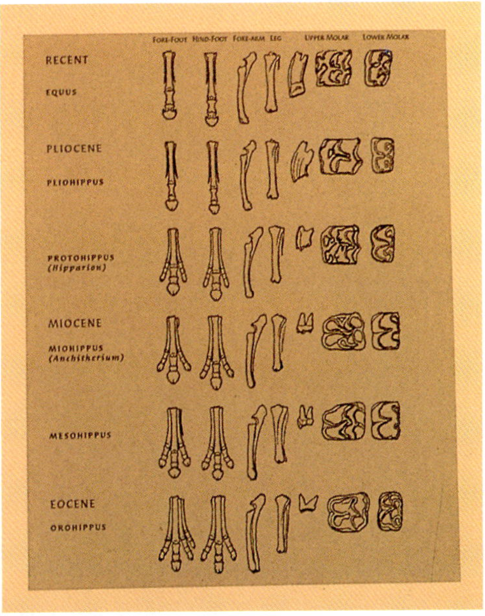

2.12 O.C. Marsh's (1877) genealogy of the main line of evolution in the horse during the Tertiary showing the gradual reduction of the side toes on the fore and hind legs.

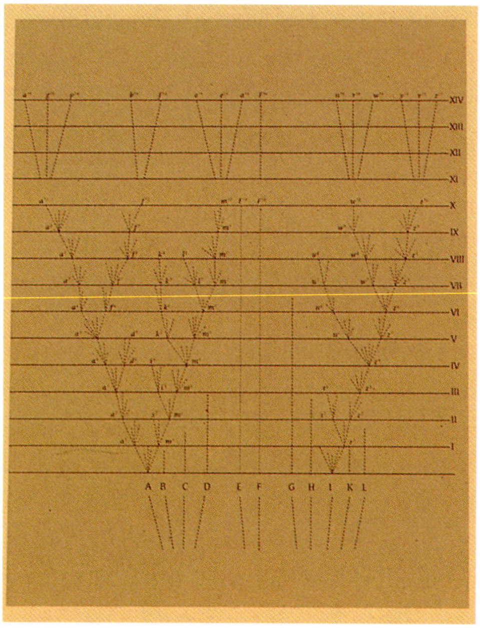

2.13 Although Darwin himself adopted the kinship form in his diagram of evolution, he has certainly inspired other scientists of his time to use 'family-trees' (1859).

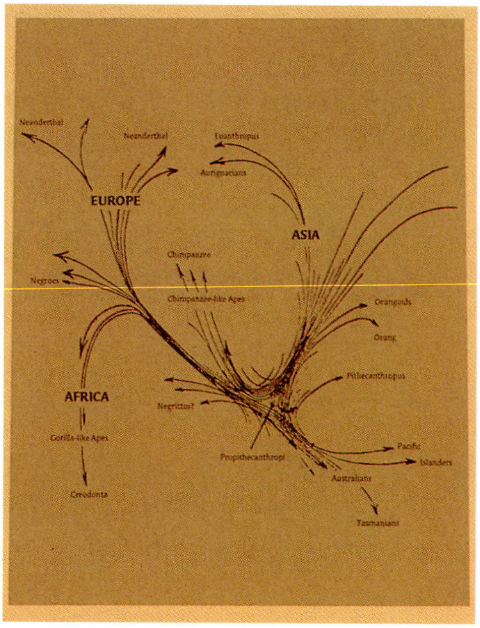

2.14 The evolutionary scheme of the anthropologist Hermann Klaatsch (1913) showing the evolution of primates and various human races.

ric and contemporary times had initiated the study of ecosystems since the early nineteenth century. Buckland (1836) developed a fold-out diagrammatic section of the earth's crust, showing the typical rock formations and their characteristic fossils.

Analogous to the abovementioned efforts to classify humans as part of the primate order, presented in his later book, *Antropogenie oder Entwicklungsgeschichte des Menschen,* Haeckel(1876) presented a reconstruction of the human genealogical tree, based on comparative embryological information. His rather misleading dogma of 'ontogeny recapitulates phylogeny', postulated that every individual must go through the entire evolutionary process of its species during its development. In this 'family-tree', it is suggested progression proceeded through twenty-two stages from the most 'simple' organisms (*Moneren*) via amphibians (*Amphibien*) up to modern humans (*Menschen).* As an illustration, he also drew an evolutionary chart of the 'Modern Theory of the Descent of Man' tracing back the progression of all organisms from protoplasm to Papuan.

According to Haeckel, humankind had evolved during the Tertiary Period out of a primate-line encompassing an Ape (*Pithecos*) as progenitor to Man (*Anthropos*), with an Ape-Man (*Pithecanthropus*) in between as a transitional creature. This Ape-Man constituted the 'link' between Man and Ape, the ultimate connection between Man and Nature to be reconstructed from the past. As he placed the *Pithecanthropus alalus* (speechless Ape-Man) between the twenty-second stage of modern humans and the twentieth stage of the apes, these creatures

2.15 *Genealogy from Abraham, Isaac, and Jacob's sons to Herron, Judah's grandson.*

2.16 *The lineage of Adam, the son of God. First part of Bendorp's seven part 'Tree of Jesse' showing the ancestry of Christ.*

2.17 *The Royal lineage from Judah to David, showing the succession of David's sons.*

2.18 *The genealogy from Noah to Abraham, showing the early inhabitants of the world after the Flood.*

were perceived - despite the lack of articulated speech - as having a human habitus. Evolutionary trees were introduced by Haeckel in the 1860s. They became the standard iconography for evolution.

It was Darwin, however, who drew upon the imagery of kinship to describe the relations among the living species and to delineate human ancestry. The use of the tree form for portraying possible ancestral relationships between Man and Ape - phylogenies - had been inspired by the use of 'family-trees' in historical genealogy. It resembled the biblical 'family-tree' of Jesse depicting the earthly ancestry of Christ.

This form of portraying prehistoric human ancestry equally relates to the concept of the forest, which as a symbolic aspect of the natural world had brought ecology into evolutionary thinking of the nineteenth century. This early tradition is also expressed in elaborate heraldic 'family-trees' of royal houses and aristocratic families, often decorated with various related family coats of arms. In this way, family histories are expressed, while kinship relations can also be traced back from different coats of arms. Genealogical descriptions are neither restricted to the past nor to West-European culture. Famous are the Moslim genealogies, very well delineated by calligraphers.

Contrary to expectations, the role of the different human fossils found during the second half of the nineteenth century in the ongoing debate of human descent during the same period of time was rather limited, as Bert Theunissen(1989) well documents. Despite important finds such as of fossil skullcap of Neanderthal Man by Johann Carl Fuhlcott near Düsseldorf in 1885, the Neanderthal jawbone by Edouard Dupont near Dinant in 1866, and Cro-Magnon Man in 1868 in the Dordogne, questions of evolutionary theory and human descent were scarcely raised in this context. While Huxley (1863) assumed that these fossils belonged to a primitive form of the human race and Darwin regarded them as human, the pathologist Robert Virchow argued that the bones had been deformed by disease, rickets and arthritis in particular.

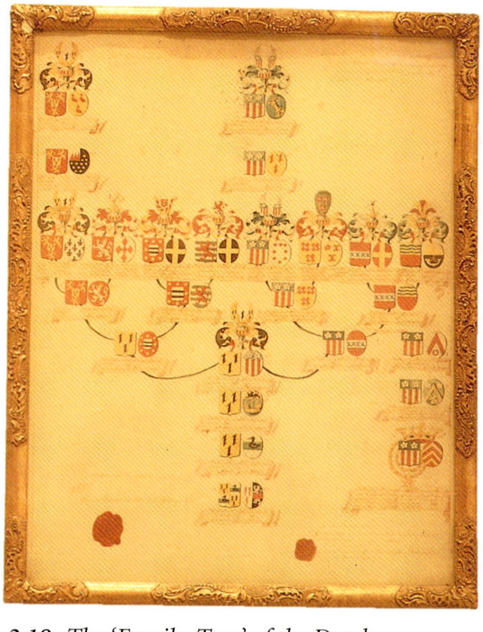

2.19 The 'Family-Tree' of the Dutch Schuilenborg Family.

It is noteworthy that even Haeckel, who had helped to establish the German School of Morphology that had begun to concentrate on the study of human descent, ignored the Neanderthal fossils for a long time. The main interest of this School of Thought was in the reconstruction of *phylogeny,* the evolutionary history of life, but instead of pursuing palaeontology, comparative anatomy and embryology formed its principal objectives. No special research efforts were yet undertaken in the search for human fossils or missing links, presuming discoveries often too fragmentary and inconclusive.

2.20 Coat of Arms of Wilhelmina Glimmer, wife of Bertram van Eck van Panthaleon, Lord of Groenewoude and Oud-Broeckhuysen. This seventeenth-century coat of arms shows the simplified ancestry of a Dutch aristocratic family.

2.21 'Family-Trees' have continued to attract our attention. A modern version by O. Blunder (1982) shows the 'One and Only Tree of Blunders'.

2.22 Two Moslim genealogies from the Library of Vienna, Austria.

The related field of ethnology - later evolved into anthropology - was equally interested in the origin of modern human races, but in common with their morphologist colleagues, they showed hardly any interest in human fossils. The question of human origin was approached from either a monogenist or a polygenist point of view. While the first view regarded all human races as having had the same origin, the latter assumed each race evolved separately. Although the idea of the 'Great Antiquity of Man' was generally accepted, the genealogical relationships between Modern and Prehistoric Man barely attracted any attention. Only much later, with the development of palaeoanthropology as a sub-discipline of general anthropology, did the explanation of human origins become methodologically structured, identified with 'key events' in the process of human evolution, *i.e.* encephalization; terrestriality; bipedality; and culture.

In short, most nineteenth-century scientists engaged in the study of human's place in nature started to focus on comparative anatomical research of man and apes, on embryology, and - only yet in a very modest way - on fossil evidence of early ancestors.

Against this background of such major shifts in the Study of Man during the second half of the nineteenth century, unprecedented in the history of science, Dubois nurtured his ambition to study the development of human polygenetic evolution, and go back further in time in order to substantiate the relationship between man and nature.

3.0 *Dubois' library: a window on other worlds.*

3 Dubois' Passion for the 'Missing Link'

Born on 28 January, 1859, - a year before the publication of Darwin's famous *Origin of Species* - in the small village Eysden on the banks of the River Maas, Eugène Dubois very early developed a keen interest in the study of the natural world.

As he admitted later in his life, it was the splendid natural landscape of South Limburg in particular with its many rich resources that from his childhood strongly influenced the direction of his scholarly destination: "My birthplace and first home... undoubtedly sowed the seeds of the naturalist in me, or at least developed them".

His father Jean Joseph Balthasar Dubois (born on 15 June, 1832), who had come from Thimister, a small town in Belgium near the Dutch/German border as a boy of six, later became the Burgomaster of Eysden. Following his apprenticeship in Eckelrade, by the age of 20, Jean had already obtained his diploma as 'rural pharmacist', and the same year established a pharmacy in the new family house in the *Breusterstraat* in Eysden.

3.1 Eugène Dubois (left) and his brother Victor as toddlers in Eysden.

Married on 30 April, 1857, in Gronsveld to Marie Catherine Floribert Agnes Roebroek (born on 4 July, 1839) from the nearby town of Eckelrade, he founded a devout Roman Catholic family in which Eugène was the first child, followed by a brother, Jean Marie Victor Guillaume, and two sisters, Marie Antoinette Helène and Marie Jeanette Gerardine. His younger brother, Victor, later became a well-known physician in the town of Venlo, while his sister Antoinette entered the Religious Order of

3.2 The house of the Dubois Family in Eysden where Eugène was born (current state 1992).

the Ursuline Sisters as Mère Marie-Angélique. His sister Gerardine married Dr. Gerard Fortemps, who practised medicine in the nearby Belgium town Fouron-le-Compte, just across the border.

The scientific interest of young Eugène was soon further encouraged by his father, who taught his son the Latin names of the local animals, plants, trees, and shrubs. Later on, Eugène often helped his father to collect medicinal plants and herbs for use in the pharmacy. His special attention was drawn to the unique small hill with an elevation of 360 feet near Eysden, the 'Sint Pietersberg'. While rich humus on its top has provided local farmers with abundant fertile soil for generations, its underlying high-terrace gravel deposit covers Mesozoic limestone chalk or *mergel* (marl), that has been used as building material for houses and cellars for centuries. Soon after the easily accessible marl deposits became depleted, builders had to tunnel deep into the mountain to reach their supply.

Since the Middle Ages an ever-growing, extensive network of subterranean tunnels, galleries, and corridors has established a vast labyrinth, in which the young student collected his fossils and studied rare plants found only in limestone habitats. He knew all twelve species of bats which use the underground caves and passages during the winter.

Eager to learn more about the world in which he grew up, he plunged into the study of the natural history of Southern Limburg to which he would return later in his career: "From my boyhood bedroom the continuation of the St. Pietersberg into Belgium (was) visible over the (river)Maas. I walked there many a time to collect fossils, and to the sand of the S(outh) L(imburg) chalk plateau close to the eastern boundary of Eysden."

Inspired by natural history books such as Buffon's *Histoire Naturelle* (1749-1788),

3.3 In this family picture of 1886, Eugène is standing behind his sister Antoinette who had joined the Ursuline Sisters as Mère Marie-Angélique.

3.4 Wax seal of the Coat of Arms of the Dubois Family, with its device 'Recte et Fortiter' still in use by some of its members.

3.5 The rural town of Eysden on the River Maas in South Limburg, where Eugène Dubois spend his childhood.

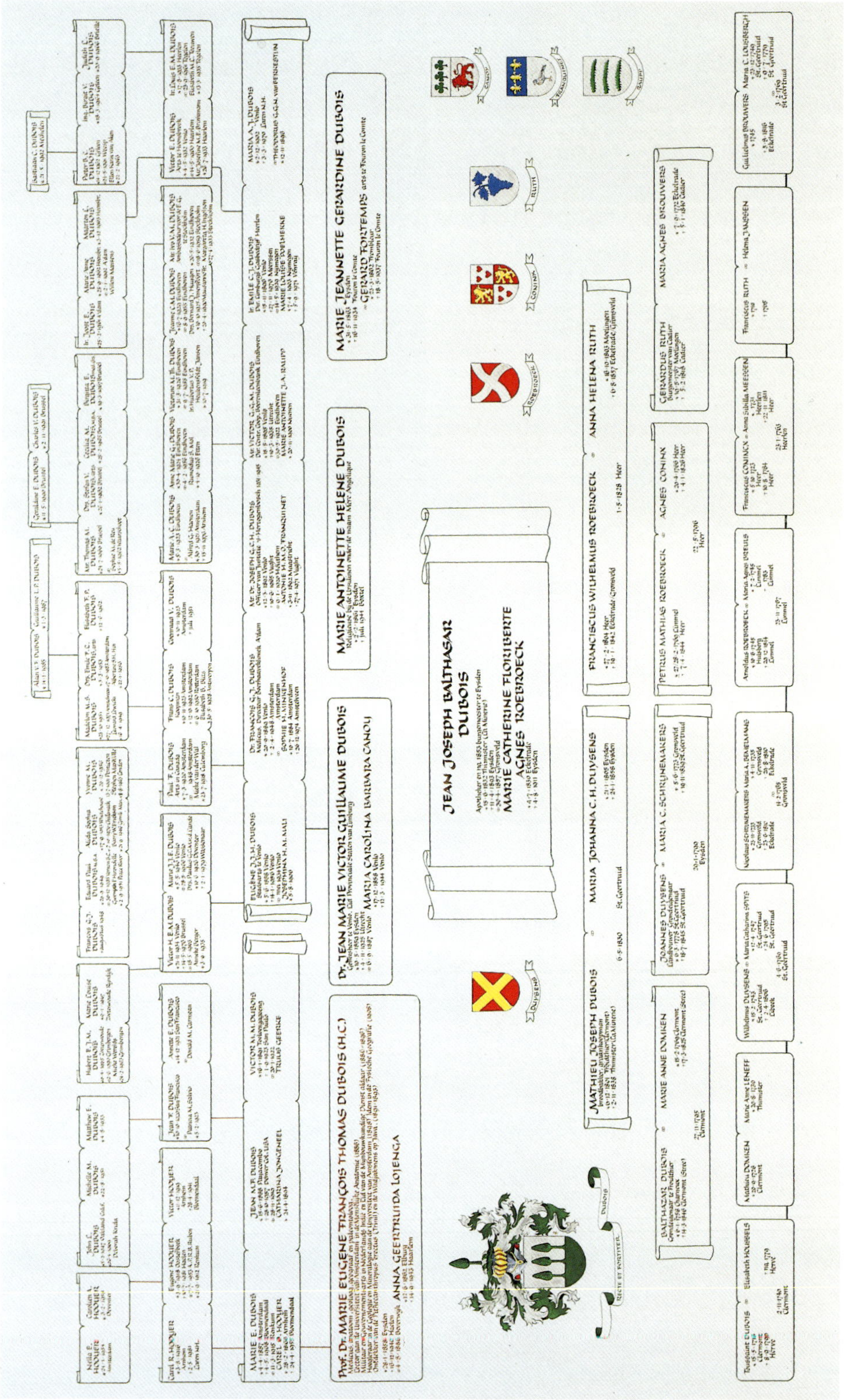

3.6 *Genealogical Quarters of the Dubois Family with the family's coat of arms and device 'Recte et Fortiter'. Designed and executed by Mrs. T. Demoed, Zutphen, The Netherlands.*

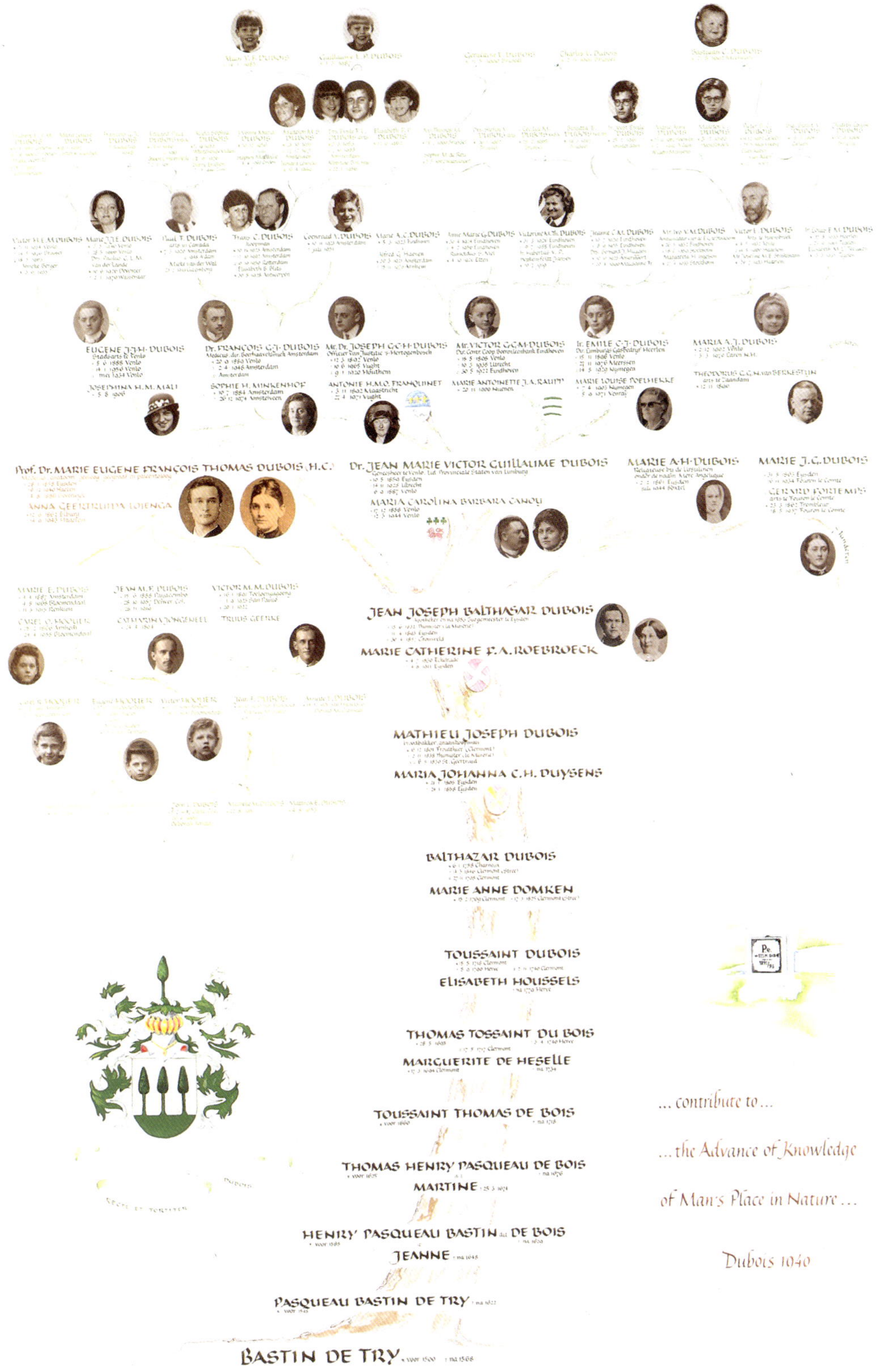

3.7 'Family-Tree' of the Dubois Family indicating Eugène Dubois' central position and the kinship relationships with his ancestors and relatives. Several portraits of the family members have also been included. Designed and executed by Mrs. T. Demoed, Zutphen, The Netherlands.

BESCHOUWING
DER WONDEREN GODS
IN DE MINSTGEACHTTE SCHEPZELEN.
OF
NEDERLANDSCHE
INSECTEN,
In hunne aanmerkelyke Huishouding, wonderbaare Gedaantewisseling en andere wetenswaardige Byzonderheeden,

Volgens eigen Ondervinding beschreeven, naar 't Leven naauwkeurig geteekend, in 't Koper gebragt en gecoloreerd

door

CHRISTIAAN SEPP.

EERSTE DEELS EERSTE STUK,

behelzende de Verhandelingen over de

DAG-VLINDERS
van de EERSTE BENDE.

Te AMSTERDAM,
Gedrukt voor den AUTEUR.
MDCCLXII.

3.8 Frontispiece of Christiaan Sepp's 'Beschouwing der Wonderen Gods in de Minstgeachtte Schepzelen of Nederlandsche Insecten' (1762).

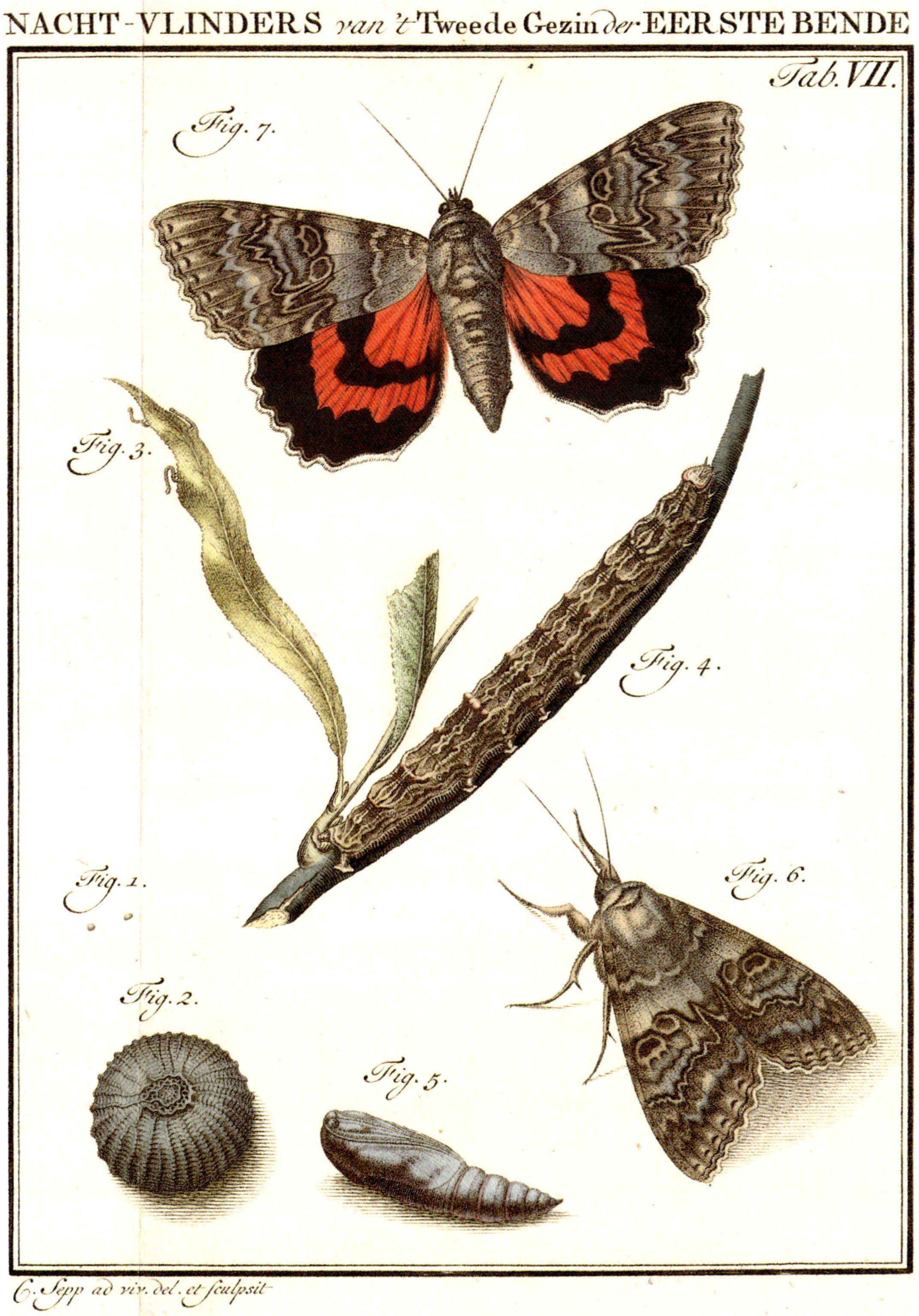

3.9 *Illustration of the red underwing butterfly from Christiaan Sepp's 'Beschouwing der Wonderen Gods in de Minstgeachtte Schepzelen of Nederlandsche Insecten' (1762).*

3.10 G.H. Schubert's (1882) illustration of different species of bats including the Flying Fox (Pteropus edulis), the Vampire Bat (Vespertilio vampyrus), and Long-ear Bat (Vespertilio auritus).

3.11 Drawer of a Dutch eighteenth-century 'Rariteiten Cabinet' filled with animal material for medicinal use (Central Museum, Utrecht).

Suringar's *Zakflora* (1870), and Calwer's *Käferbuch* (1858), he often went out wandering through the fields and woods, or along the banks of the Maas, where he started to collect plants, stones, shells, animal skulls, and insects for his own private collection of interesting natural objects. He took a special interest in the bats, which lived in the tunnels of the nearby 'Sint Pietersberg'. They would play an important role in his later cephalization research and his ecological studies in South Limburg.

Since the new geographical discoveries of the 17th century, the Dutch Golden Age, collecting the 'wonders of nature' had become a mania for many people, inspired by increased

3.12 'Anatomisch Amphitheater' of Leiden University with various skeletons and objects originating from the former 'Rariteiten Cabinet', used for scientific investigations. Copper engraving by Bartholomeus Dolendo (1609).

contacts with new, unknown parts of the globe. Indeed, as the cargoes of the *Vereenigde Oost-Indische Compagnie, VOC* (The United East India Company), ships brought a host of souvernirs, curiosa, and ethnic artefacts from distant parts of the world to Amsterdam, the 'warehouse of the world', collecting became an important medium to seize upon new knowledge and understand the rapidly-expanding world. In contrast to other countries, where collectors consisted mainly of aristocrats and princes, collecting in Holland had become a middle-class passion, in which doctors, lawyers, and merchants indulged in a wide-ranging establishment of a microcosm, a complete world in miniature.

Usually kept and displayed as a whole in a special *cabinet,* collections of these objects of

3.13 *Anatomical study of the female body, Leiden, 1613.*

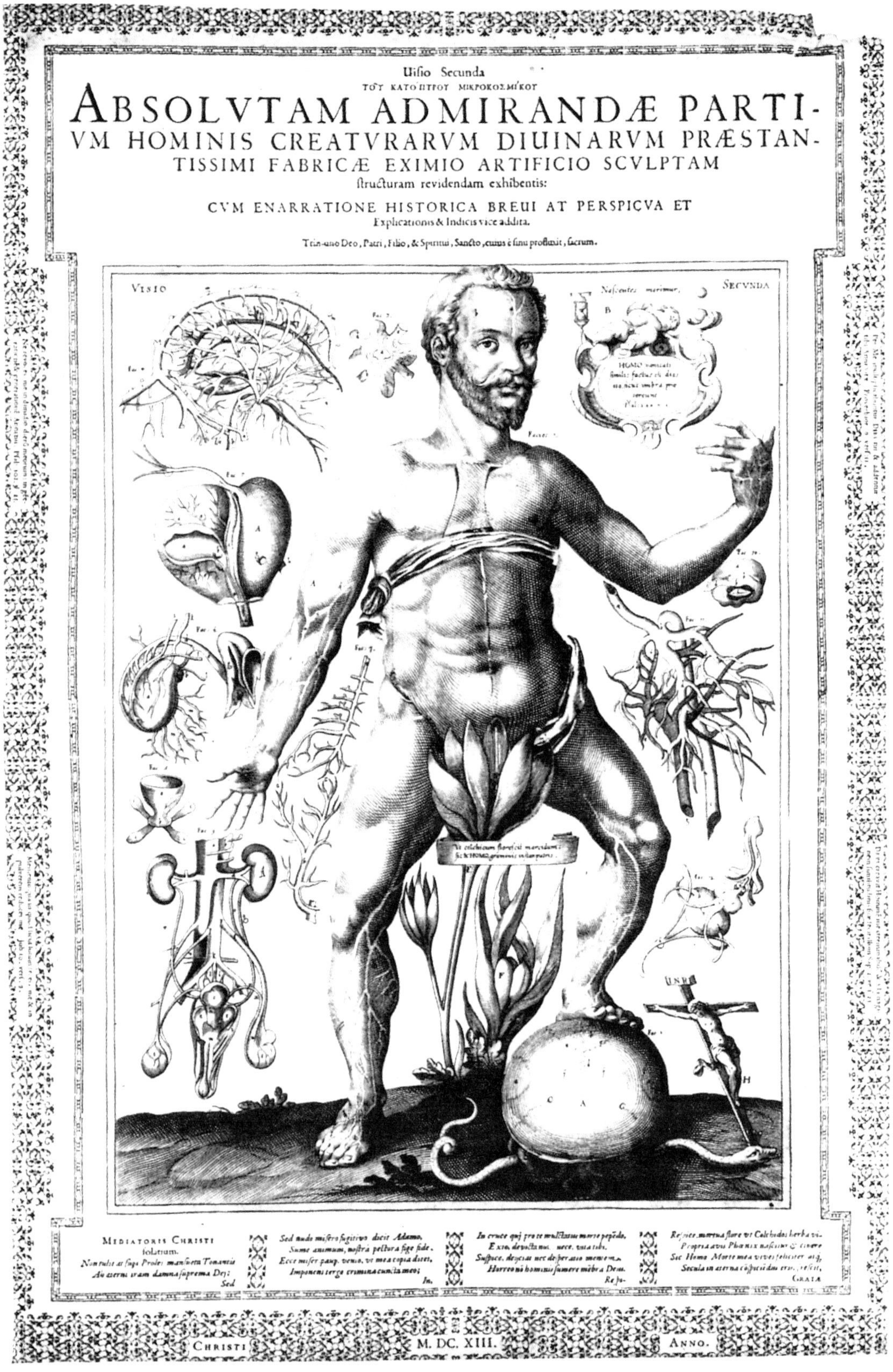

3.14 Anatomical study of the male body, Leiden, 1613.

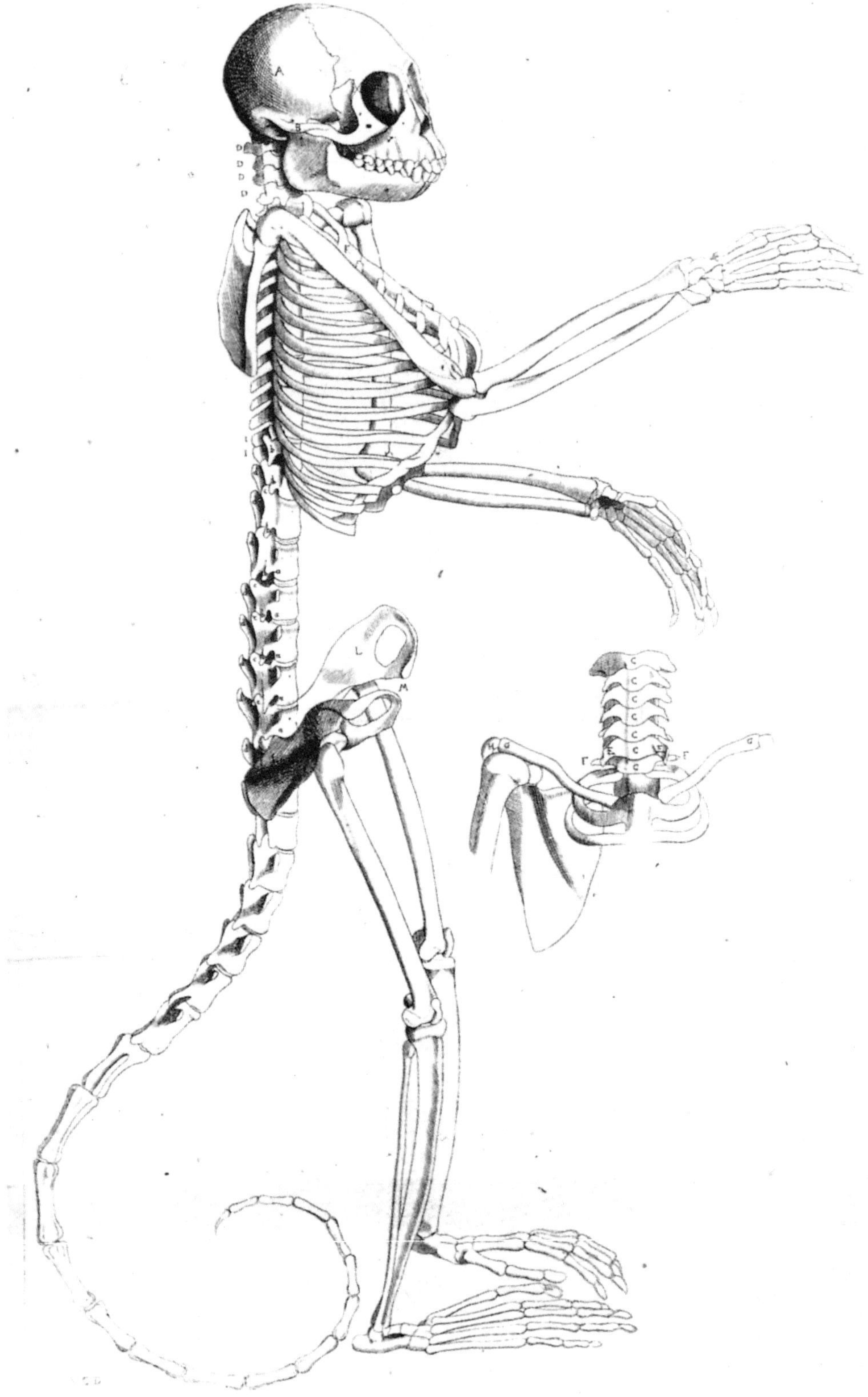

3.15 Late sixteenth-century engraving of a skeleton of a long-tailed monkey, Leiden.

general curiosity served equally as a medium for education and for scientific research. In this way, in Leiden, *i.e.* the famous *Rariteiten Cabinet* (collection of curiosities) played a crucial role in the academic endeavours of the early days of Leiden University after its foundation in 1575. Also in Leiden, initial scientific investigation in medicine focused their attention on anatomy and morphology.

Since his return from The Netherlands East Indies, the National Museum of Natural History in Leiden has housed the complete *Collectie Dubois* which includes over 40,000 objects. Recently Dubois was designated as one of the eight most remarkable collectors of the world, who, in the words of Stephen Jay Gould (1992)"... may rest assured that they have been weighted in the balance and found worthy."

It is not surprising that later in his professional career Dubois would assume the curatorship of one of the most famous museums collections in The Netherlands, the Teylers Museum in Haarlem.

On 5 September, 1870, at an age of 12, he left home to start his secondary school education at the, *Rijks Hoogere Burger School, H.B.S.,* (Dutch State High School) in Roermond. Each term he lodged with the Knarren family until he took his diploma in 1877.

It was during this time at high school that his teachers Dr. Joseph Rosen (natural history), Dr. W.H. Julius (geology and mineralogy), and Dr. J.M.A. Kramps (mathematics) further stimulated Dubois to pursue a career in the natural sciences:"...the HBS in Roermond...was probably more important to me than the university...something in the nature of a small university for me...blessed with superb teachers and just as well equipped as the universities of that time, (with) a first-rate chemistry laboratory, a large zoological, mineralogical and petrographic collection, (and) a fine botanical garden."

Around this time a German scientist, Karl Vogt, made a lecture-tour through Holland in defence of the new Darwinean theory of evolution. Although there was considerable interest in his presentations, he also met with strong opposition, especially in the predominantly Roman Catholic south of Holland, where a religiously-based creationist view prevailed. At one of Vogt's lectures in Roermond, which Dubois attended, a man in the audience rose and asked aggressively: "Do apes have churches? Do they have libraries? " which gave the young high school student a lot of amusement.

The subject of Man's Place in Nature, expounded at the time by Darwin, Huxley, and Haeckel, fascinated Dubois immensely and he followed all discussions and publications on the 'Descent of Man' closely. The question of human evolution exercised many minds, those of scientists and laymen alike, and coming from a strict Roman Catholic background, young Dubois was initially torn between evolutionist and creationist views, characteristic of the late Victorian Age.

On 18 September, 1877, he left for Amsterdam to commence his studies - despite his father's wish for him to become a pharmacist - in Medicine at the University of Amsterdam.

Much appreciated by his fellow students as a most promising young man, Dubois was easily accepted into the midst of a circle of prominent Roman Catholics of that time, including the publisher Alberdingk Thijm, the architect Cuypers, and the priest Brouwers. However, despite these Catholic connections, Dubois now turned his back on the church for ever. Later in his life, in 1932, he confided to a friend that he had abandoned religion much earlier: "... since my 13th year I have lapsed from the Catholic Church and have never again been under its influence for a single minute, right up to the present, neither in sentiment nor in deed." Throughout his life and work Dubois, as an agnostic, adhered to the principle that science and faith should be kept apart.

It soon became clear that the young student preferred to delve into the study of nature, turning away from the idea of becoming a practising physician. During his first year in medicine, which was largely made up of the study of the natural sciences, as a brilliant student he soon attracted the attention of the prominent professors of their time such as Van der Waals, Van 't Hoff, and De Vries.

He became increasingly interested in physiology and morphology, encouraged by his teachers Thomas Place and Max Fürbringer. Although in his heart more interested in physiology, later, in 1888, he accepted an assistantship in anatomy from Prof. Fürbringer, who himself had been trained at the German School of Morphology of Haeckel and Gegenbauer in Jena. Even if it later on became clear that Dubois would have much preferred to specialize more in physiology under Prof. Place, as Fürbringer's assistant he completely dedicated his work to comparative anatomy. In 1881, Dubois was appointed as Instructor in Human Anatomy at the *Rijksnormaalschool voor Tekenonderwijzers* (State Training College for Teachers of Art), and a year later, he also became an instructor at the *Rijks Kunstnijverheidsschool* (State School for Applied Arts).

3.16 Photograph of Eugène Dubois as Instructor in Human Anatomy at the 'Rijksnormaalschool voor Tekenonderwijzers' (State College for Teachers of Art) in Amsterdam of 1883.

Finally, on 16 July, 1884, Dubois passed his final examination for Doctor of Medicine, but as he had previously studied at the *Hoogere Burgerschool* instead of at a *Gymnasium*, he only received the title of *Arts* (Physician). Later, in 1897, however, after he had returned from The Netherlands East Indies, the University of Amsterdam bestowed upon him the degree of Doctor of Botany and Zoology *Honoris Causa,* the very first scholar to receive such an honour. Meanwhile, in 1885, Dubois had declined an offer to accept a Readership in Anatomy from Utrecht University, initially expecting to take over his own teacher, Fürbringer's, position in Amsterdam later on.

Around 1885 - upon the suggestion of Fürbringer - Dubois conducted an important study of the comparative anatomy of the larynx in vertebrates, which he completed in line with the conventional research methodology of phylogenetic morphology. Based on comparative anatomical research of the larynx in different animal species, he wanted to contribute to the evolutionary history of the organ. He published a summary of his results under the title 'Zur Morphology des Larinx' in the *Anatomischer Anzeiger* in 1886.

All this time, however, Dubois' interest in the study of human evolution not only from a predominantly medical, *i.e.* anatomical and physiological, but also from a palaeontological, perspective had not abated one jot.

In fact, his ambition to devote much more of his time and effort to the continuing study of the history of human evolution gradually turned into an obsession.

The only fossils of human ancestors that had been discovered at the time came first from a limestone quarry in the Neanderthal Valley in Germany , and later in Spy, in Belgium. They were those of the Neanderthals, a relatively modern form of human who became extinct some thirty-four thousand years ago. Despite the fact, that Dubois was more interested in a much earlier human form, more primitive than Neanderthal, he took a special interest in the type of natural shelter in which the Neanderthals were found: a cave sixty feet above the Düssel, a small tributary of the Rhine.

During the vacations in 1886 and 1887, Dubois undertook fieldwork in the Neolithic Flint Industry in Limburg, near his hometown Eysden. With the owner of the land, Count De Geloes, he investigated a geological 'organ pipe' near Ryckholt, called *"De Henkeput"*, where he searched for animal and human remains.

Although the German zoologist Haeckel, like most evolutionists of his time, including Darwin, Huxley, and Wallace, ignored the Neanderthal fossils, he was a great inspiration to Dubois in his anthropological passion to bridge the gap between Man and Ape. Basing himself on comparative anatomy, Haeckel had concluded that since humans and apes were somehow related, there must have been an intermediate form he called "Ape-like Man", or *Pithecanthropus*. This creature, according to Haeckel, would have evolved on *Lemuria*, a hypothetical continent thought to have disappeared beneath the Indian Ocean. The idea intrigued Dubois immensely.

In 1886, he finished his contribution on the larynx in whales for Max Weber's *Beitrag zur Frage nach dem Ursprung der Cetaceen* and started to make preparations for a comprehensive publication of the results of his larynx research, complete with drawings.

But then, suddenly, in that same year, as he later often remarked Dubois faced: "...the great turning point in his life..." Despite the fact that his future at the University of Amsterdam looked most promising as he was about to become the successor to Professor Fürbringer, he advised the Faculty of his firm determination to forego that position and find a way to go to the tropics in search of the 'Missing Link'.

Among his reasons for abandoning his academic career were his dislike of teaching, his growing dissociation with laboratory work in anatomy, and the paternalistic attitude of

3.17 Photograph of Eugène Dubois and his wife Anna Lojenga just before their departure to The Netherlands East Indies in 1887.

Fürbringer towards Dubois' independent research, which, after an incident involving copyrights to his publication, eventually led to a definite farewell to anatomy.

As Theunissen (1989) rightly notes, by taking this significant step he changed the course of conventional nineteenth-century research in human evolution to emphasize the crucial role of material evidence from the fossil record.

His decision fell like a thunderbolt among his colleagues: Fürbringer and Weber tried everything in their power to make him change his mind, considering his plan utter folly, even comparing him with *Hans im Glück* in Grimm's *Fairy Tales.* Dubois was ridiculed as the foolish boy who bartered away all his posessions for something he believed better, only to lose everything in the end.

All this was in vain. Shortly afterwards, he resigned from the university, in his farewell lecture sharing his ambitious plans to undertake an expedition to the tropics in search of the long-sought 'Missing Link' with his students.

On 29 October, 1887, with his wife Anna and infant daughter Marie Eugènie, he set out for The Netherlands East Indies on board the steamship *Prinses Amalia.*

3.18 On 29 October, 1887, the steamer 'S.S. Prinses Amalia' took the Dubois family to the island of Sumatra in the former Netherlands East Indies.

3.19 Dubois (standing in the middle) and his wife Anna (below right) with their fellow-passengers on board of the steamer 'S.S. Prinses Amalia' on their way to The Netherlands East Indies.

4.1 Map of Eastern Java.

4. Lemuria: Journey to The Netherlands East Indies

After leaving the university, Dubois spent a great deal of time and energy to mount his expedition to the East Indies. He quickly realized that a private enterprise would be out of the question. He then decided to approach the Department of Colonies in The Hague. After listening to his plans with some amusement, the Secretary-General refused him financial support, advising him to forget all about the nonsense he had been reading in Darwin's 'absurd' book. Although Dubois had half expected such a reaction, stemming from its tiny chance of success, the controversial character of the research, and the unfavourable economic climate at the time, he became even more determined than ever to find his way to the East Indies.

Why specifically to the East Indies? The answer to this, for him such a vital question, is largely answered in an article on the *Desirability of Research on the Diluvial Fauna of the Dutch Indies, Especially of Sumatra* that he later published in the *Natuurkundig Tijdschrift* (Journal of Natural History) in 1888.

In addition to his general theoretical considerations, derived from well-known views of nineteenth-century evolutionists such as Darwin and Haeckel, he was the first to introduce a new approach which would provide tangible proof of our direct relationship with the natural world. This proof, he believed, was to be found in the past, in the vicinity of the lost continent of *Lemuria*. Man's closest evolutionary relatives such as the early primates and prosimians, including the lemur, the ancestor of the various contemporary lemurs of Madagascar, and the *mouse lemur*, must have lived on this continent.

Indeed, in his *Descent of Man*, Darwin (1871), supposed that human ancestors must have lived in the warmer climate of the tropics since humans had lost their pelts of hair in the course of evolution. Equally, since human's closest relatives in the animal kingdom, the apes, were living in the tropical regions of the globe, early common ancestors could also be expected to have lived there as well.

The fact that a few years earlier, in 1878, Richard Lydekker had discovered fossil remains of a primate - the Siwalik Chimpanzee - in British India only strengthened Dubois' belief in the East Indies as the possible birthplace of humanity. This, however, was in contrast to Darwin's view that Africa has been the 'Cradle of Mankind'. As we now know after more than half a

b
b
a

4.2 G.H. Schubert's (1882) illustration of prosimians including the Howler Monkey (Mycetes ursinus), the Capuchin Monkey (Cebus capucinus), the Lion-Monkey (Hapale rosalia), and the Lemur (Lemur).

century of extensive work on *Australopithecus, Homo habilis,* and, most spectacularly, on the well preserved early *Homo erectus* - the 'Turkana Boy' - in the African context, the 'pithecanthropine' *i.e.* Far Eastern examples of *Homo erectus,* are most probably derived from African ancestry.

Finally, Dubois had been intrigued by the contemporary occurence of gibbons which are closely related to humans, in the Malay Archipelago. The abundant occurrence of caves throughout the East Indies - supposedly providing natural shelter for early ancestors as they had done for European Neanderthals - only whetted Dubois' eagerness to also engage in cave research in the East Indies as soon as possible.

In the end, only one option remained, the one way for a determined Dubois to reach his goal: Holland's colonies in the area, The Netherlands East Indies. Dubois decided to do what several Dutch scientists such as the ichthyologist Pieter Bleeker and ethnologist/physician Cornelis Swaving had done some forty-five years before him. Firmly bent on going he signed up for eight years as Medical Officer, Second Class, with the rank of Second Lieutenant in the Medical Corps of the *Koninklijk Nederlandsch-Indisch Leger, KNIL* (Royal Netherlands East Indies Army).

Initially Dubois had pinned his hopes on the numerous caves in the limestone mountains of the island of Sumatra. Here, a particular mountain range in the Padang Highlands, called *Boekit Ngalau Sariboe* (Mountains of A Thousand Caves), appeared full of promise.

So, after a voyage lasting about seven weeks on the *S.S. Prinses Amalia* of the *N.V. Stoomvaartmaatschappij 'Nederland',* Dubois - who had meanwhile grown a beard - and his family arrived in Padang, an army post on the West Coast of Sumatra. The first months were very busy, Dubois being kept occupied with his hospital work. As he had hardly any time to devote to his scientific work, he soon requested to be transferred to the interior. Mainly through the influence of the Governor of the West Coast, R.C. Kroesen, and his successor W.P. Groeneveldt, he was fortunate enough to be transferred to a new post located in the Padang Highlands. In gratitude for the support of both these gentlemen at the start of his research in Sumatra, Dubois later on named two significant fossil discoveries from Trinil on Java after them: *Tetraceros Kroesenii,* an antelope, and *Felis Groeneveldii,* a tiger.

His above-mentioned article in the *Natuurkundig Tijdschrift* of 1888 was primarily intended to raise the interest of The Netherlands East Indies Government, and consequently to obtain funds for his research plans. Rather a strategic paper, it indicated favourable prospects for research into the fossil fauna of The Netherlands East Indies, only mentioning the possibility of retrieving human fossils towards its end. It ended with the rhetorical question: *"And will The Netherlands, which has done so much for the natural sciences of the East Indian Colonies, remain indifferent when such important questions are concerned, while the road to their solution has been indicated?"*

From his posting at Pajakoemboeh Dubois sought to prove these assumptions and, by August 1888, in the Lida Adjer Cave he had indeed collected an impressive range of fossil remains of animals including deer, elephants, gibbons, pigs, rhinoceroses, and tapirs.

4.3 The mountain Ophir in the hinterland of West Sumatra where Dubois searched the caves for fossils. Coloured lithograph by F.C. Wilsen in De Indische Archipel 1865-1876.

His ideas soon gained the support of prominent scientists such as Melchior Treub, Chairman of the *Indisch Comité van Wetenschapelijk Onderzoek* (East Indies Committee for Scientific Research), while in Holland his erstwhile colleagues Max Weber and Karl Martin, and the members of the *Commissie tot Bevordering van het Natuurkundig Onderzoek der Nederlandsche Koloniën* (Commission for the Promotion of Research in the Natural Sciences in The Netherlands Colonies) ranged themselves behind his cause.

Consequently his report on these promising results in faunal research was brought to the attention of The Netherlands East Indies Government, and in conjunction with the ideas expressed in his previous article, his proposal was finally accepted.

By Government Order No. 6 of 6 March, 1889, Dubois was exempted from military service and assigned to the *Departement van Onderwijs, Eeredienst en Nijverheid* (Department of Education, Religion, and Industry), the Director of which, W.P. Groeneveldt instructed Dubois to continue his scientific work in Sumatra, and later on in Java. Two staff members of the military Engineering Corps, Franke and Van den Nesse, joined his team which was completed by a group of fifty forced labourers.

After this initial success he extended his research area over the whole of the highlands, but soon grew disappointed with the follow-up results. As he wrote in a letter of 17 October, 1889, to F.A. Jentink, Director of the *Rijksmuseum voor Natuurlijke Historie* (National Museum of Natural History) in Leiden: "Everything here has gone against me, and even with the utmost effort on my part I have not achieved one hundredth part of what I had visualized."

4.4 'Life-size' illustration of the skull of the 'Wadjak Man' found by Van Rietschoten in the Wadjak Cave in Java in 1888.

Not only had his expeditions into the Sumatran hinterland proved perilous in terms of the many accidents which befell his staff and labourers, frequent attacks of malaria, lack of water, and largely inaccessible research sites, but - even worse - the scientific results fell far short of Dubois' expectations. In essence, the fossil fauna of Sumatra did not differ from contemporary species rendering the likelihood of the presence of a transitional form in the area very small.

Towards the end of 1889 Dubois decided to shift his research efforts to the island of Java, which seemed more promising than Sumatra in many respects. While the geologist Verbeek had already referred to the promising Tertiary limestone in Java, an unusual find of a fossil human skull by the mining engineer B.D. van Rietschoten near the East Javanese village Wadjak on 24 October, 1888, turned the scales. C.P. Sluiter, the Curator of the *Koninklijke Natuurkundige Vereeniging* (Royal East Indies Society for the Natural Sciences) in Batavia (Jakarta since Independence), who had received the skull from Van Rietschoten, passed it on to Dubois for his opinion. Dubois was very exited about the discovery, which was proof of the existence of fossilized human remains on the island. After close examination, he distinguished the Wadjak skull strictly from those from the contemporary inhabitants: "I am virtually certain that the first representative of the primordial people of Java has now been discovered."

After receiving permission from the Government, published in a Resolution of 14 April, 1890, to extend his research to the island of Java, Dubois and his family, including a newborn son, Marie François, left to take up the new assignment.

4.5 *Dubois sorted out his fossils on the veranda of his house.*

4.6 Dubois' friend Boyd, a Scots planter who lived on the Wilisea plantation in East Java, introduced the Dubois family to the research area of Toeloeng Agoeng.

After careful consultation, he selected the mid-eastern part of Java as the area for his research activities. He settled in Toeloeng Agoeng, a small town in the *Residentie* (District) of Kediri, which would serve as his headquarters for more than five years. The Dubois family moved into a large one-storey house with spacious galleries, quite an improvement compared to the previous army quarters in Sumatra. While normally such a front gallery would be reserved for resting or receiving guests, it gradually became an excellent place to store all kinds of fossils, packed and unpacked.

North of Toeloeng Agoeng rises the extinct volcano Mount Wilis, its fertile slopes then covered with coffee plantations. The manager of one such estate, Wilisea, was a Scotsman named Boyd who soon became Dubois' close friend, sharing his love of Nature. During the regular visits of the family - now extended by a newborn son, Victor Marie Mathieu - to the peaceful estate, the elderly gentleman provided the newcomers with valuable information on the region in which he had lived for so many years. Learning about Dubois ambition to find the 'Missing Link', he warned him to have patience, as things could not be rushed in Java.

Dubois used his time well, exploring the region with its rivers and forests and its special geological formations. Later on, his examination and classification of various strata would prove most valuable in prospecting for fossil-bearing locations. His first expeditions were made into the area where Van Rietschoten had previously discovered the Wadjak skull: the limestone mountains of the Kediri and Madioen Residencies. After the beginning of a successful range of excavations producing fossil remains of mammals and reptiles now extinct on the island, the research was extended into Van Rietschoten's rockshelter.

It was here that at the end of September 1890, a second, less complete, skull was discovered pertaining to the same Wadjak Man. Dubois attributed the scientific name *Homo wajakensis*

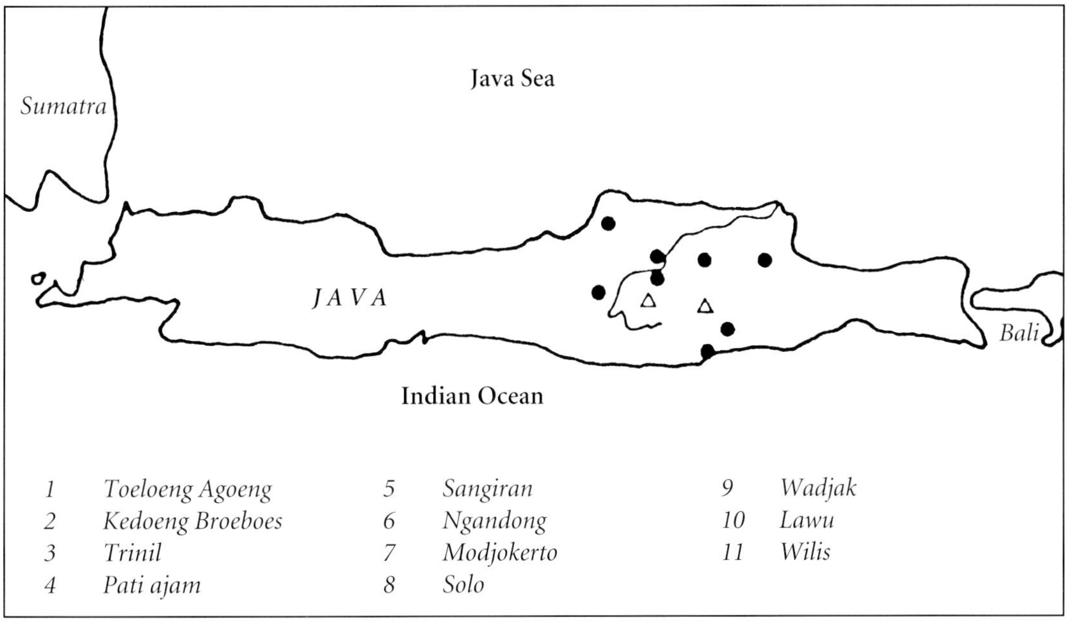

4.7 *The various excavation sites of Dubois in East Java, according to Theunissen, 1989.*

to this race. His handwritten progress report of November 1890 on the find to the Director of the *Departement van Onderwijs, Eeredienst en Nijverheid* mentions that: "These Australian inhabitants of Java undoubtedly lived in a time one could call geologically recent", confirming that he considered that Wadjak Man was directly related to modern humans.

Previous explorers, such as the famous Javanese scholar and painter Raden Saleh and the German naturalist Franz Wilhelm Junghuhn, had earlier collected many fossils on the slopes of the Kendeng Hills in Central and East Java and on Pati Ajam Mountain in the *Residentie* of Japara. Since the majority of these fossils were collected over a large area, outside caves and rockshelters, Dubois decided also to start excavating in the open countryside.

His team now included two sergeants from the Engineering Corps, G. Kriele and A. de Winter, each of whom had the supervision of half the group of labourers. Field excavation was conducted in the Kendeng Hills, notably in Kedoeng Loemboe and Kedoeng Broeboes.

Dubois assigned De Winter to collect fossils and look for fossil-bearing deposits in the region of Pati Ajam Mountain. However, De Winter's expedition met with several set-backs. The volcanic slopes were covered with tough tall-growing grass, *alang alang*, rendering accessibility to the surface extremely difficult. Surprisingly, his expedition also met with unexpected competition from the local people who had been digging for fossils in order to sell these to Chinese merchants. Indeed, the traditional Chinese pharmacopoeia includes the use of powders and potions made of ground fossil bones and, for centuries, Chinese have sold a variety of such 'dragon bones' (*Lung-ku*) as medicine.

This practice was not only met with during later fieldstudies in China by anthropologists in search of the 'Peking Man', such as Walter Granger, Roy Chapman Andrews, and J.G. Anderson, but has also been observed in the larger cities outside China proper. Ralph von

4.8 *Junghuhn's bivouac near the top of the mount Gedeh in Java, during his volcanological studies in the mid-1850s.*

4.9 *River view in East Java around 1842.*

Koningswald, for example, visited the Chinese pharmacies in Batavia (Jakarta) and Hongkong where he purchased various fossil teeth, ready to be ground into toothache cures. In this way Von Koningswald was able to acquire some unusual fossil teeth in a Hongkong pharmacy. Close examination revealed them to be the teeth of a giant ape from the Middle Pleistocene which he named *Gigantopithecus.*

Initially De Winter was less fortunate since the Chinese refused to sell. Gradually, however, he managed to buy a number of fossils from some Javanese, but they declined to disclose their find-spots.

Upon De Winter's request to the local authorities for assistance, an order was soon issued in the villages that no more fossils were to be dug up and sold. Nevertheless, it proved to be too profitable a business to relinquish. The trade continued clandestinely, and several of the excavated fossils were even stolen from the actual excavation sites. Eventually, the Pati Ajam expedition, which had found fewer fossils than expected, was discontinued, and De Winter and his team returned to Madioen in August 1891.

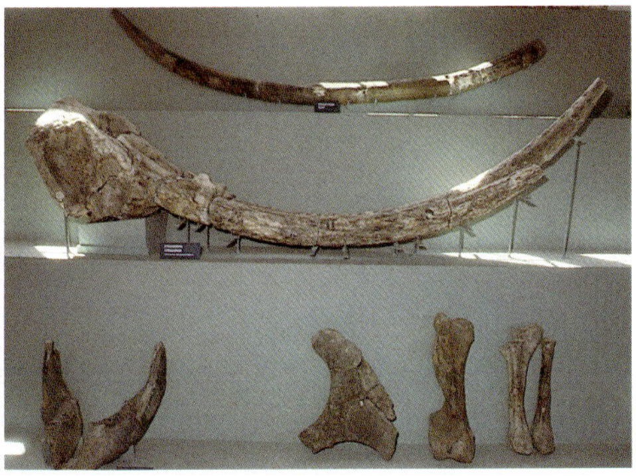

4.10 Fossil remains of a stegondon (Stegondon trigonocephalus) from Trinil, East Java. (Collectie Dubois).

4.12 Fossil remains of a hippopotamus (Hexaprotodon sivalensis) from Trinil, East Java. (Collectie Dubois).

4.11 Molar of an elephant-like animal from the Ci Saat fauna at Trinil, East Java.(Collectie Dubois).

4.13 Fossil remains of a new species of prehistoric antilope to which Dubois' own name was given: Duboisia santeng from Trinil, East Java. (Collectie Dubois).

4.14 Fossil remains of rhinoceros (Rhinoceros sondaicus), tapir (Tapirus indicus), orang-utan (Pongo pymaeus), bear (Ursus malayensis), and Indian elephant (Elephas maximus) from Trinil, East Java. (Collectie Dubois).

4.15 Fossil remains of a pig (Sus Brachyonathus) and a tiger (Panthera tigris) found from Trinil, East Java. (Collectie Dubois).

4.16 Fossil remains of a pig (Sus macrognathus), hyena (Hyaena brevirostris), pangolin (Manis palaeojavanica), and a molar of a primitive elephant from the Kedoeng Broeboes fauna fromTrinil East Java. (Collectie Dubois).

4.17 Fossil remains of barking deer (Muntiakus muntjak), deer (Cervide), and a hominid from Trinil, East Java. (Collectie Dubois).

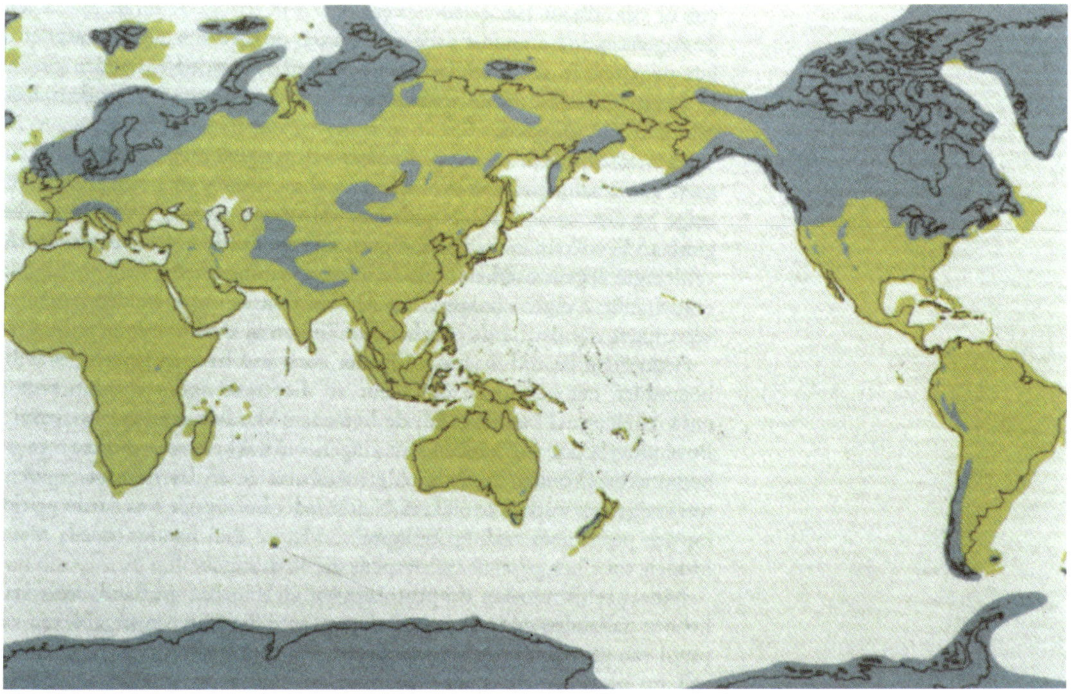

4.18 Shallow seas around Indonesia which dried up during the glacial periods.

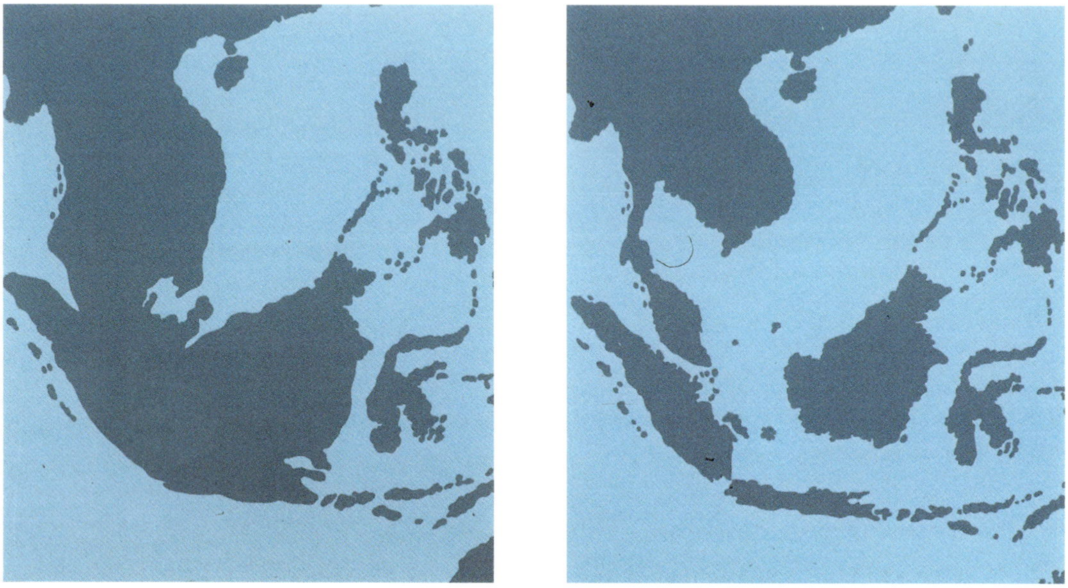

4.19 During the glacial periods, the island of Java was regularly connected with the Asian mainland.

Dubois himself, assisted by Kriele, had been investigating the area of the valley of the Solo River (*Bengawan Solo),* and many fossils were soon being discovered in its banks. The strata of the banks consisted entirely of cemented volcanic tuff, clay, sand, and lapilli stone. In Dubois' opinion, they showed a great similarity to the formations in the Siwalik Hills of Northern India.

In order to visualize the prehistoric environment of early humans in the area, Dubois systematically collected the faunal and floral fossils of the different strata.

Among the fossils of the region were important mammal remains including a primitive elephant (*Stegodon*) and hippopotamus (*Hexaprotodon sivalensis*). Also, a new species of prehistoric antilope was unearthed, to which Dubois' own name was given: *Duboisia santeng.*

In addition, fossil remains were found of rhinoceros *(Rhinoceros sondaicus*), tapir (*Tapirus indicus*), orang-utan (*Pongo pymaeus*), bear (*Ursus malayensis*), and Indian elephant (*Elephas maximus*). Furthermore, fossils were collected of pig (*Sus macrognathus*), tiger (*Panthera tigris*), hyena (*Hyaena brevirostris*) and pangolin (*Manis palaeojavanica*) as well as of barking deer (*Muntiakus muntjak*), deer (*Cervide*), and a hominid.

During the glacial periods, the Indonesian islands were regularly connected with the mainland as a result of falling sea levels. This enabled animals from South-East Asia to migrate and spread over Java, Sumatra, and Borneo, and eventually led to the appearance of *Homo erectus* in Java. The climatic changes influenced animal life, pertaining to subsequent changes in the fauna in Java.

John De Vos and Paul Sondaar recently developed a scheme of seven different fauna found in Java, based on the fossil record. These include:

1 The *Satir* fauna showing that about 1.5 million years ago Java was an island, on which deer, hippopotamus and an elephant-like animal with cusped molars lived.
2 The *Ci Saat* fauna contained predators and another elephant-like animal (*Stegodon*), indicating a connection between Java and the mainland 1.2 million years ago.
3 The *Trinil* fauna of one million years ago, when the connection with the mainland was even better. During this period, *Homo erectus* appeared in Java.
4 The *Kedoeng Broeboes* fauna witnessed a new wave of migration about 800,000 years ago, in which a modern type of elephant is found: *Elephas hysudrindicus.*
5 The *Ngandong* fauna includes the rhinoceros (*Rhinoceros sondaicus*) and a modern type of elephant (*Elephas hysudrindicus).*
6 The *Punung* fauna of 80,000 years ago contains orang-utan (*Pongo pygmaeus*), gibbon (*Simia hylobates*), and the modern elephant (*Elephas maximus*).
7 The *Wadjak* fauna of 10,000 years ago is identical to the contemporary Javanese fauna.

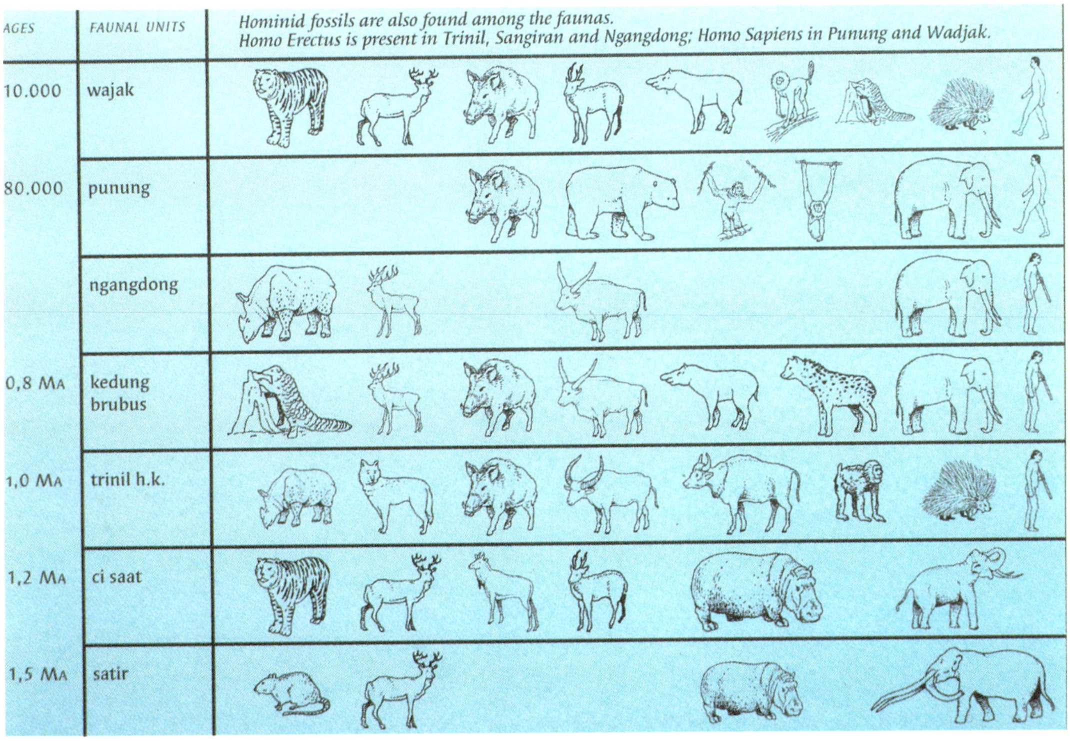

4.20 *Scheme of seven different fauna found in Java, reconstructed from the fossil record in the Dubois Collectie in Leiden.*

All in all, the results of both expeditions were extremely rich in fossils. Since 1891, more than 12,000 fossils had been found at the Trinil site, prompting Dubois to classify several specimens as originating from the *Javanese Siwalik* fauna comparable to British-Indian Siwalik. Most remarkable was the discovery of a number of bones of the giant pangolin (*Manis*), three times the size of the existing Javanese species (*Manis javanica Desm.).* Dubois attributed the name *Manis palaeojavanica* to this prehistoric pangolin.

The excavations were carried out fairly accurately, using a 'grid-system' in which significant fossils were marked with the number of the grid in which they were found. The altitude was measured from the lowest water level.

In view of the overwhelming results of the excavations near the Solo River, Dubois decided to concentrate his future work in this promising area. Here he was to work in the coming period between 1891-1893.

4.21 Dubois' own photograph of the excavation site on the banks of the Solo River near Trinil, East Java.

4.22 The Trinil Museum near Trinil, East Java during its official opening in November, 1991.

5.1 Eugène Dubois back in Holland from The Netherlands East Indies at the age of 40 (October 1902).

5 Disclosure and Debate: Man-Ape or Ape-Man

It was in the year 1893 that Dubois finally decided to officially disclose his discovery of the 'Missing Link' to the world. He sent a telegram to his friends in Europe in which he confidently announced that he had found the 'Missing Link of Darwin'.

In the same year, at the end of the dry season, a small monument indicating the find of *Pithecanthropus erectus* was placed near Trinil, on the right bank of the Solo River opposite the excavation site. Its inscription reads: "P.e. ⟵ 175 M. ONO —≪ 1891/93" indicating the geographical direction and its exact distance from the find-spot.

Before the famous discovery of the skullcap in September 1891, an unusual discovery consisting of a molar had been made on the same site at Trinil. Dubois described it initially as: "... a molar (the third molar of the upper right side) of a chimpanzee (*Anthropopithecus*). This genus of anthropoids, occurring only in West- and Central Equatorial Africa today, lived in British India in the Pliocene and, as we can see from this discovery, during the Pleistocene in Java."

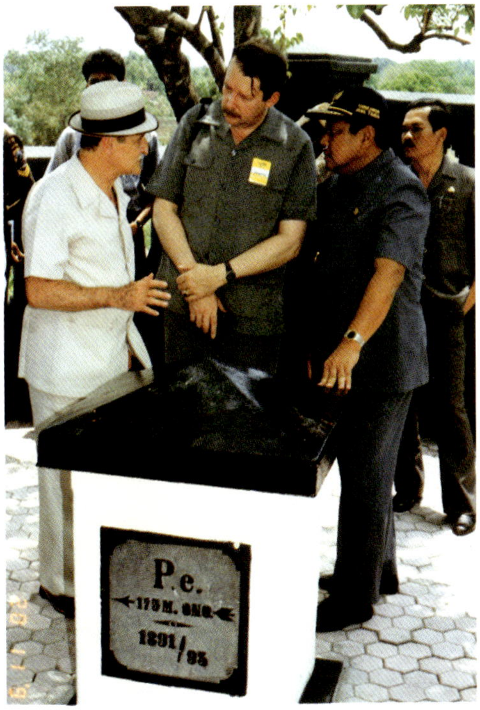

5.2 Prof. Phillip Tobias discussing Dubois' discovery with Mr. Sularso, the Governor of East Java and Dr. Jan Slikkerveer at the P. e. monument near Trinil (November 1991).

The famous skullcap itself - found in October 1891 - was ascribed to the same anthropoid ape: "Close to the place on the left river bank where the molar was found, a splendid skullcap was dug up which, just as unquestionably as (the) molar, has to be attributed to the genus Anthropopithecus (*Troglodytes*)."

Clearly, in this context, he used Lydekker's genus name *Anthropopithecus*

5.3 Delegates to the International Trinil Centennial Colloquium in Surabaya crossing the Solo River on their way to the find-spot of Pithecanthropus (November 1991).

5.4 The excavation site on the banks of the Solo River around the time of the discovery of the Pithecanthropus fossils, photographed by Dubois (around 1893).

emphasizing the direct relationship between his Javanese chimpanzee and Lydekker's Siwalik chimpanzee from British India.

In his letter of August 1892 to the Director of the Department, Dubois expressed his conviction that the Trinil fossils belonged to a human-like ape. He refers to *Anthropopithecus javanensis* , the 'Man-Ape from Java'.

Interrupted by the rising waters in the wake of the monsoon, the work at Trinil could only be resumed in May 1892, and in August a third spectacular 'anthropoid' fossil was discovered in the bank of the Solo River: an almost complete left thigh-bone. In his subsequent report in "*Palaeontologische Onderzoekingen*" (Palaeontological Researches) of 1893 he describes its position: "This thigh-bone lay at the same level at which the other parts were found, yet following the direction of the earlier stream which deposited the tufa material 15 m. upstream."

By now Dubois was convinced that all three fossils belonged to one and the same individual, and this last discovery of the femur had finally settled the matter. Indeed, as the fossil thigh-bone showed such a strong resemblance to a human thigh-bone, he concluded that his *Anthropopithecus* from Java stood upright and moved like a human. In his opinion this creature was in no way equipped to climb trees as chimpanzees, gorillas, or orang-utans are. Consequently, Dubois went a step further, extending the original name of his creature to *Anthropopithecus erectus.* The 'upright-walking Man-Ape' in 1892.

Enthousiastically he wrote in 1892: "So the first intermediate form has been made known in this Lower Pleistocene anthropoid from our island which, in an unmistakable way, links Man more closely to his next-of-kin among the mammals."

Remarkably, however, in his last report of 1893, Dubois has changed the initial name *Anthropopithecus* (Man-Ape) to *Pithecanthropus* (Ape-Man). In the *Collectie Dubois* in Leiden, there is a manuscript by Dubois, in which the fossils are first described as *Anthropopithecus*, the genus name, which later on has been crossed out and replaced by *Pithecanthropus.*

So, it was in the year 1893 that he realized that he had found a transitional species Pithecanthropus which would go down in history as his 'Missing Link'. Consequently, this dramatic moment was chosen for the celebration in 1993 of the 'Pithecanthropus Centennial 1893-1993'. Obviously this change of name was mainly inspired by the more human characteristics of the thigh-bone, the form of which was closer to that of Man than to that of Ape.

Finally, his discovery had become what it still is today: the 'Ape-Man from Java'.

Now that his early ancestor fully answered the *Pithecanthropus* postulated by Haeckel and, as a representative of the twentieth stage in the human 'family-tree', must have originated in the sunken continent of *Lemuria*, in 1893 Dubois started to prepare his official publication. As a follow-up to his previous articles, it would encompass a comprehensive treatise on his discovery. Although circumstances in The Netherlands East Indies were far from ideal for completing an in-depth study of the fossils due to lack of reference and comparative materials, he managed to produce a description of the fossils before he actually left Java.

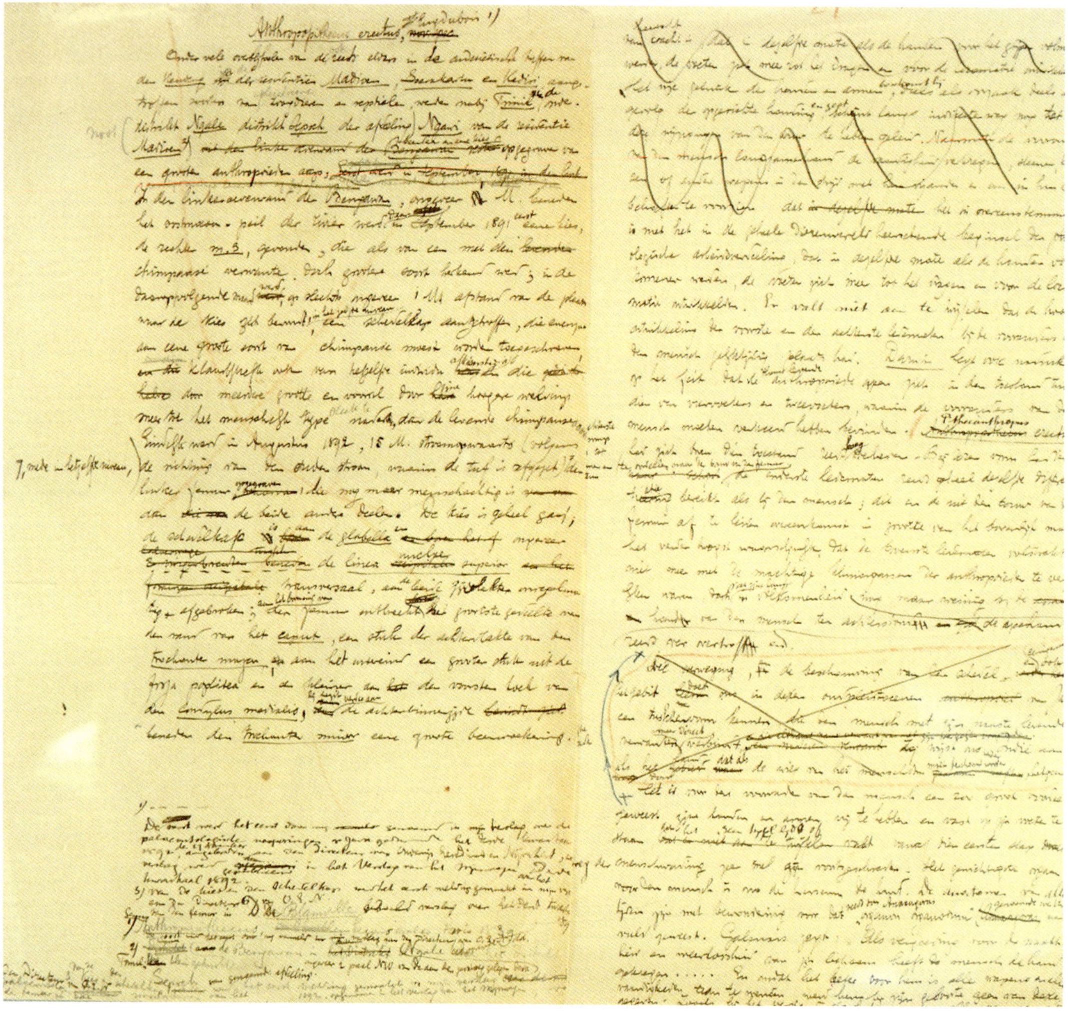
Anthropopithecus erectus, Eug. Dubois

5.5 In 1893, Dubois after realizing that he had found the 'Missing Link' between Man and Ape changed the name 'Anthropopithecus' (Man-Ape) to 'Pithecanthropus'(Ape-Man).

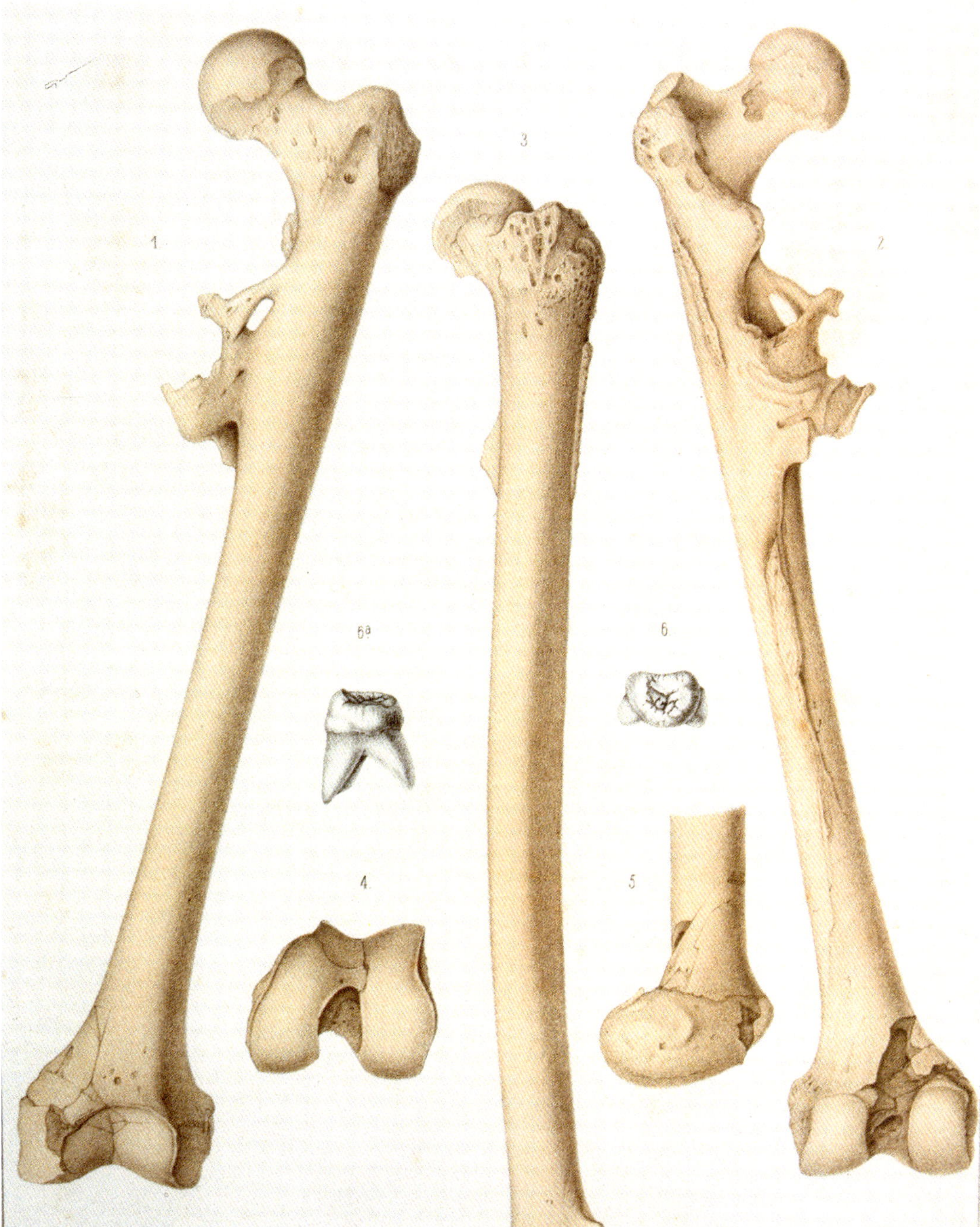

5.6 *Dubois' drawing of the Pithecanthropus thigh-bone and molar (Lithograph 1894).*

5.7 The contemporary occurence of the closely to humans related gibbon (Simia hylobates) in the Malay Archipelago induced Dubois to go to South-East Asia.

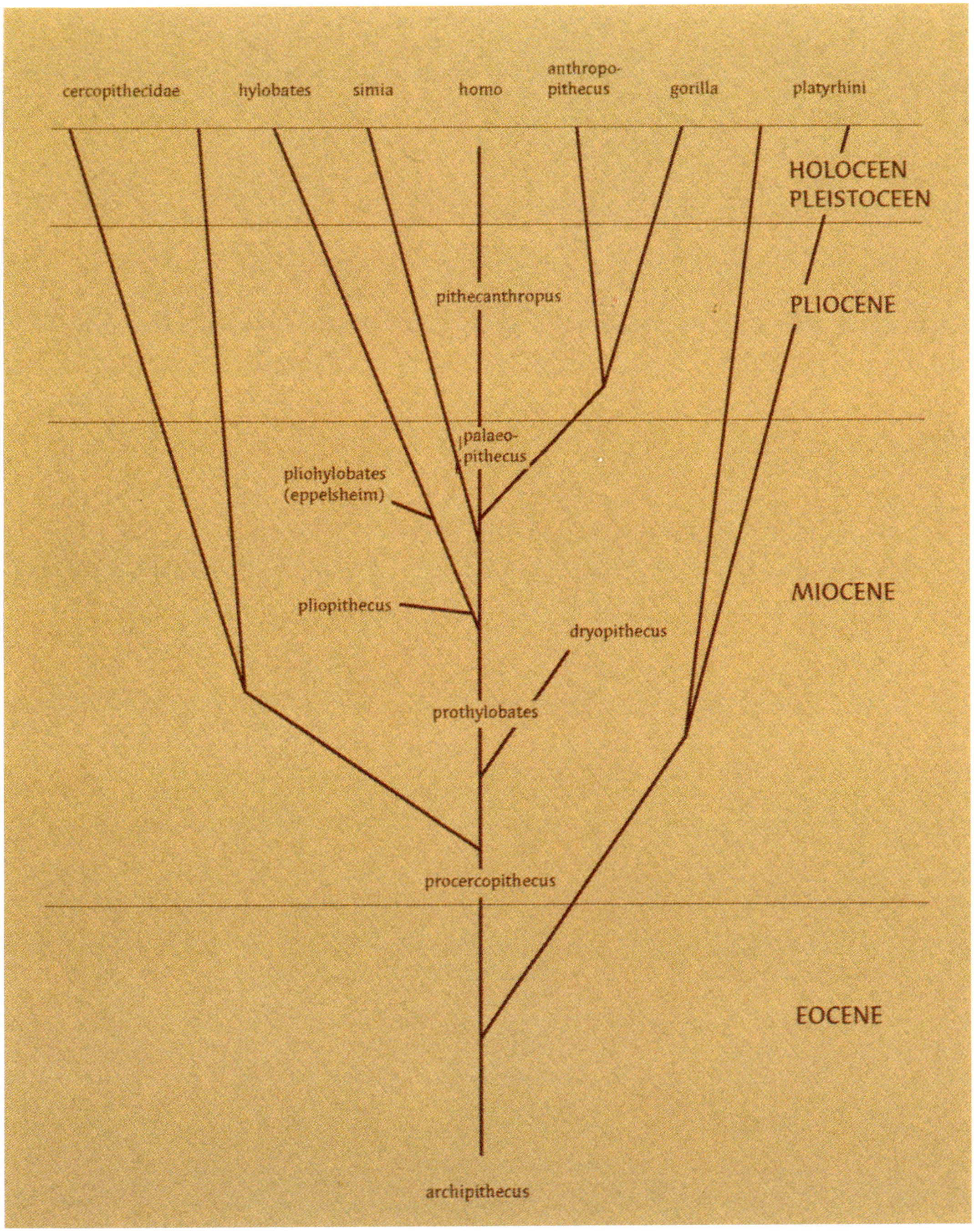

5.8 Dubois' phylogenetic 'family-tree' showing Pithecanthropus as the 'Missing Link' between Man and Ape (1896).

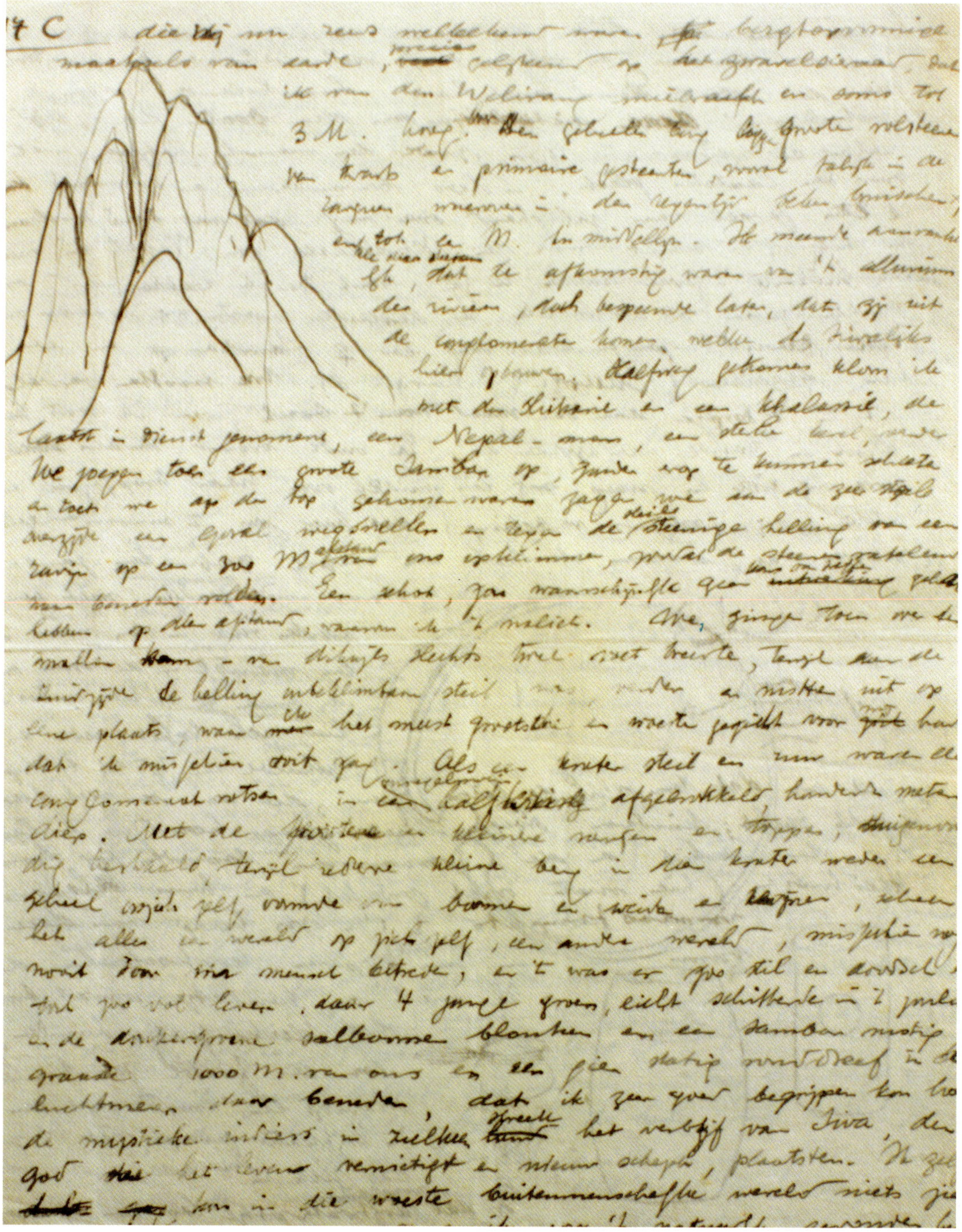

5.9 *Page of Dubois' diary of his expedition to the Siwalik Hills in India, with a sketch of the find-spot of the rare molar of the Stegolophodon stegodontoides (1895).*

The next year, Dubois' famous publication: *Pithecanthropus erectus, eine menschenaehnliche Uebergangsform* came out in Batavia (Jakarta). The 39 page work starts with a description of the species *Pithecanthropus*. Then the history of the discovery is presented, followed by an anatomical description of skullcap, molar, and thigh-bone. The publication, well-illustrated with several photographs and drawings by the author himself, ends with some general conclusions. In his effort to establish the phylogenetic position of his early ancestor in the human 'family-tree', he left the description of the circumstances, the location, and the methodology of excavation rather incomplete.

In his conclusions, he confirms that, as an transitional form, the Ape-Man stood between the genus *Homo* and the anthropoids. As a precursor of *Pithecanthropus*, he postulated Lydekker's *Anthropopithecus sivalensis* from British India, a creature which, in turn would have stemmed from a gibbon-like ancestor, *Prothylobates*. Gibbons are found exclusively in South-East Asia.

Thus, Dubois was convinced that humankind had evolved from these early ancestors in the so-called 'paradise' of Haeckel's *Lemuria*. His interpretation of the human 'family-tree', his *Phylogenetic Tree of Apes and Man*, would include *Archipithecus - Procercopithecus - Prothylobates - Anthropopithecus sivalensis - Pithecanthropus erectus - Homo sapiens*, in which his Ape-Man stood between Man-Ape and Man. Not surprisingly, he agreed with Haeckel, who - contrary to Darwin's theory - had supposed that Asia had been the geographical location of human origins.

In the meantime, it had become clear that Dubois would not extend his contract as Medical Officer with the *Koninklijk Nederlandsch-Indisch Leger*, and he decided to return to Holland. Before his departure in the summer of 1895, he made a trip to British India in order to study the Siwalik fauna and flora. He visited the botanical gardens of Singapore and Calcutta, and studied the vertebrate collection in the Indian National Museum. During a fieldtrip to a number of sites into the Siwalik Hills, he collected several fossils, among which was an extremely rare molar of *Stegolophodon stegodontoides*, now part of the *Collectie Dubois* in Leiden.

His passage to British India only served to strengthen his vision about the direct relationship between Siwalik and Javanese fauna.

5.10 Extremely rare fossil molars of elephant-like animals Archidiskodon planifrons (left) and Stegolophodon stegodontoides (right) which Dubois discovered in the Siwalik Hills of British India (1895).

Upon his return to The Netherlands in early August 1895, Dubois was confronted with considerable criticism, which had been accumulating since his description of the fossils had become available in Europe. This criticism was mainly focused on the exact location of the discovery and the supposed relationship between

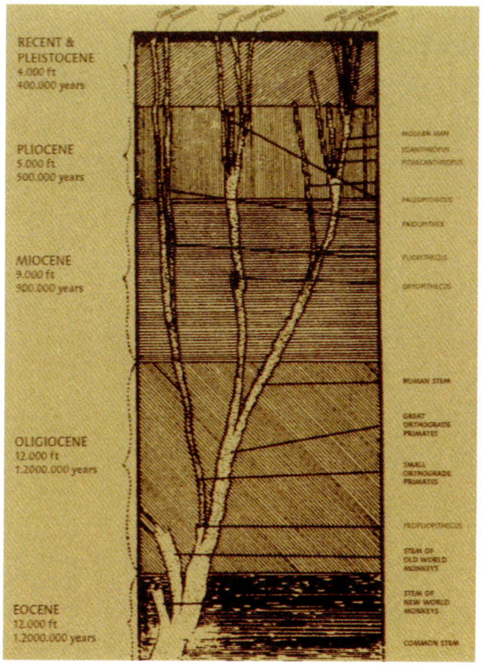

5.11 *Arthur Keith's phylogeny of the primates (1915).*

the remains, while opinions also widely differed on the interpretation of both the skullcap and thigh-bone. In general, the views of the international scientific community varied from Ape to Man, with only an occasional attribution to an intermediate form of either Man-Ape or Ape-Man.

The French anthropologist Leonce Manouvrier, who later on would become one of Dubois' leading champions, initially reacted with caution. In view of the lack of details on the find-spots, he considered the hypothesis that all three fossils belonged to one individual not proven. Despite his doubts about the status of the creature, in 1895 he noted during a meeting of the *Societé d'Anthropologie* that Dubois' discovery was nonetheless very important since it embodied the existence of transitional forms between Man and Ape, as predicted by the theory of evolution. In his view, the fossils were fairly unhuman, more ape-like. The reaction from the French zoologist August Petit was equally careful, but he tended to attribute the fossils to a human.

The German reactions to Dubois' 'Missing Link' as a transitional form were in general very negative. The anatomist Wilhelm Krause's attack concentrated on the skullcap and the molar, both of which he attributed without a shadow of a doubt to an ape. In his opinion, the thigh-bone was human. The prominent anthropologist/physician Rudolf Virchow also ascribed the skullcap without hesitation to an ape, probably a large gibbon. He concluded that the deformation (*exotosis*) of the thigh-bone had resulted from disease, which would not rule out the possibility that it also had belonged to an ape.

The renowned German morphologist Ernst Haeckel, on the other hand, fully endorsed Dubois' claim. In his publication *Systematische Phylogenie der Werbeltiere* (1895), he notes on the fossils: "Some of these are certainly of great importance, especially the skullcap of the Pliocene *Pithecanthropus erectus* of Java (1894), which really seems to represent the 'missing link' so eagerly sought for in the chain of transitional forms."

In 1896, Gustav Schwalbe, an anatomist from Strasbourg published an in-depth study of no less than 225 pages on the Dubois' fossils, in which he compared them with humans, apes and Neanderthals. In his conclusion, he accepted the idea of Pithecanthropus being an intermediate form, albeit in a slight different phylogenetic line: Pithecanthropus between the Apes and Neanderthal Man, which latter creature he regarded as a separate species, closest to Man: *Homo primigenius.*

On the whole, the British reactions were rather sceptical. Richard Lydekker, a palaeontologist, considered the fossils fully human, and supposed that the skullcap probably had belonged to a deformed idiot. The anatomist Daniel Cunningham devoted a lot of time to his critique. He reproached Dubois for only comparing the fossil skullcap with ape skulls, and in an article in *Nature* (1894-1895), concluded that it:"... is unquestionably to be regarded as human". The anatomist Arthur Keith, who was involved in primate studies, agreed with Cunningham that the Trinil fossils belonged to a man, however primitive. The anatomist William Turner, in a lecture to the Royal Society of Edinburgh on 4 February, 1895, also disagreed with Dubois. He assumed that the fossils could have come from a human race comparable to the Neanderthals. Such a possibility was also mentioned by the Swiss anthropologist Rudolf Martin who also ascribed the skullcap to the Neanderthals in his critique in *Globus* (1895).

On the American side, however, Dubois received very flattering reactions. The palaeontologist Othniel Marsh (1895) regarded Dubois' Pithecanthropus as one of the most important discoveries since that of the Neanderthal. He noted that Dubois had proved the existence of a prehistoric anthropoid, so far unknown, to science.

His Dutch colleagues also did not spare Dubois their negative reactions. The Secretary of the *Aardrijkskundig Genootschap* (Geographical Society), J.A.C.A. Timmerman, had already added a critical note about a "hasty conclusion" to Dubois' early reference to the

5.12 The group of delegates to the Third International Congress of Zoology in Leiden, photographed in the Van der Werffpark on 18 September, 1895. Dubois is in the middle of the back row.

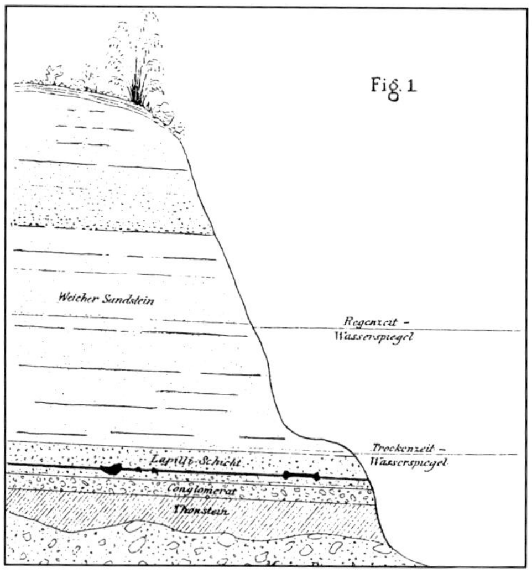

5.13 Dubois' drawing of the profile of the bank of the Solo River at the Pithecanthropus site (1895).

East Indies as the 'cradle of the human race' in his article in the Society's Journal of 1894, and this was compounded by the Dutch anthropologist Herman ten Kate who later joined in the scepticism. In his opinion, the skullcap and molar had certainly belonged to a large Ape.

Barely had he returned from The Netherlands East Indies, than Dubois had the opportunity to answer his critics personally during the Third International Congress of Zoology, held in Leiden from 16 - 21 September, 1895. On the last day of the congress he showed the Pithecanthropus fossils to his colleagues and elucidated his interpretation. His session, presided over by Rudolf Virchow, attracted much interest, and among the more than one hundred participants there were prominent scientists such as Flower, Milne Edwards, and Marsh.

Dubois began the lecture by acknowledging that in his publications certain aspects of his study had been insufficiently dealt with. He then elaborated on the issues under criticism. He presented a detailed account of the find-spot, illustrated by a profile drawing of the excavation site, followed by a well prepared defence of his interpretation of the fossils. After he had introduced a new, in his opinion related, second molar found near the skullcap location, he concluded that the various critiques had not led to any revision of his interpretation of a transitional form: Ape-Man.

The discussion that followed, ended rather indecisively. Krause (1895) reported on the Session in the *Naturwissentschaftliche Wochenschrift:* "...several misconceived opinions, to which the first publication on the subject gave rise, have certainly been removed by these discussions.....A more or less certain and definitive answer to the question, however, still lies in the future."

Following his Leiden lecture, the next year Dubois went to great lengths to visit numerous scientific institutions where he showed the fossils and defended his views: Liège, Brussels, Paris, London, Edinburgh, Dublin, Berlin, and Jena. His discovery was widely praised, but the critique of his interpretation did not fade away. Although several scientists and experts now tended to take his side, the controversy persisted.

Shortly afterwards, in 1896, he published a first reconstruction of the skull in the *Anatomischer Anzeiger.* It shows a clear blend of human and ape-like characteristics, supporting its transitional form, albeit with an unmistakable emphasis on the gibbon-like teeth and prognathism of the face. The French anthropologist Manouvrier, in a second supporting study of Dubois' Ape-Man in 1895, had compared the fossils with skulls of apes, humans, and Neanderthals. His reconstruction of the Pithecanthropus skull, that showed less ape-like features, still expresses the same mingling of ape and human characteristics.

In 1897, Dubois resumed his research on cephalization - the relative growth of the brain. In 1898, at the Fourth International Congress of Zoology in Cambridge, he presented his first results in relation with *Pithecanthropus erectus.* Later on, especially after 1918, he elaborated his hypothesis, which pertained to a new model of evolution. Based on studies of endocranial casts of a variety of animals, he developed various 'cephalization ratios' largely dependent of the number of brain cells. For animals such as the mouse, hare, and squirrel, these ratios would be 1 : 2 : 3, but for other animals, ratios of 1 : 2 : 4 were measured. As Bert Theunissen (1989) has shown, Dubois explained the regularity that stemmed from sudden increase of the brain quantity through a doubling of all or half the brain volume. Dubois expressed the abrupt leaps in the evolutionary process of mammals in a phylogenetic tree in the shape of a 'horse tail'.

In the public domain, in which his 'adventurous discovery' had already become very popular, Dubois was also active. For the World Exhibition in Paris in 1900, he made a life-size reconstruction of the Ape-Man from Java, clearly showing its position halfway between Man and Ape. The detailed reconstruction of the Pithecanthropus, with its half human, half apelike anatomy, also suggested the use of tools as an antler was placed in the Ape-Man's hand. Exhibited afterwards in the Ethnological Museum in Leiden, the statue of the Ape-Man soon gained the popular name *'Pietje'* ('Peterkin').

As the initial debate on the Ape-Man had been largely concentrated on the find-spot and the assumed relationship between the fossils, several of Dubois' critics tended to change their judgement after seeing the fossils in reality, complemented by Dubois' additional information. The number of his supporters had certainly increased, particularly in France and Germany, but a great deal of opposition to his ideas still persisted. On the whole the British maintained their human interpretation of the fossils. Rudolf Virchow, who had previously already engaged in a heated debate with Ernst Haeckel, now continued his fierce attacks on Dubois. As a champion of the German 'Ape-Thesis', Virchow soon gained support for his rejection of the existence of a 'link' between Man-Ape and Man from among the circles of German critics of evolutionary theory and the process of humanization.

On the other hand, Dubois' phylogenetic approach, that recognized the existence of a transitional form between Man and Ape to be found in the fossil record, gradually gained ground. It became increasingly apparent, that the fossil record could provide solutions to so far unsolved questions of human ancestry.

Dubois' journey to the East Indies had struck the imagination of many of his contemporaries, but initially only a few actually yet dared to engage in the search for hominid fossils in the tropics. In 1907 Margarethe Selenka mounted an expedition to Dubois' excavation site on Java under the auspices of the *Königliche Bayerische Akademie der Wissenschaften* (Royal Bavarian Academy of Sciences). This massive search for hominid fossils in Trinil involving extensive excavations yielded no further trace of Java Man; only a large number of mammal fossils was unearthed. Similarly, the well prepared expeditions, about a decade later, onto Central Asia led by Roy Chapman Andrews did not shed any new light on early human ancestry in the Far East.

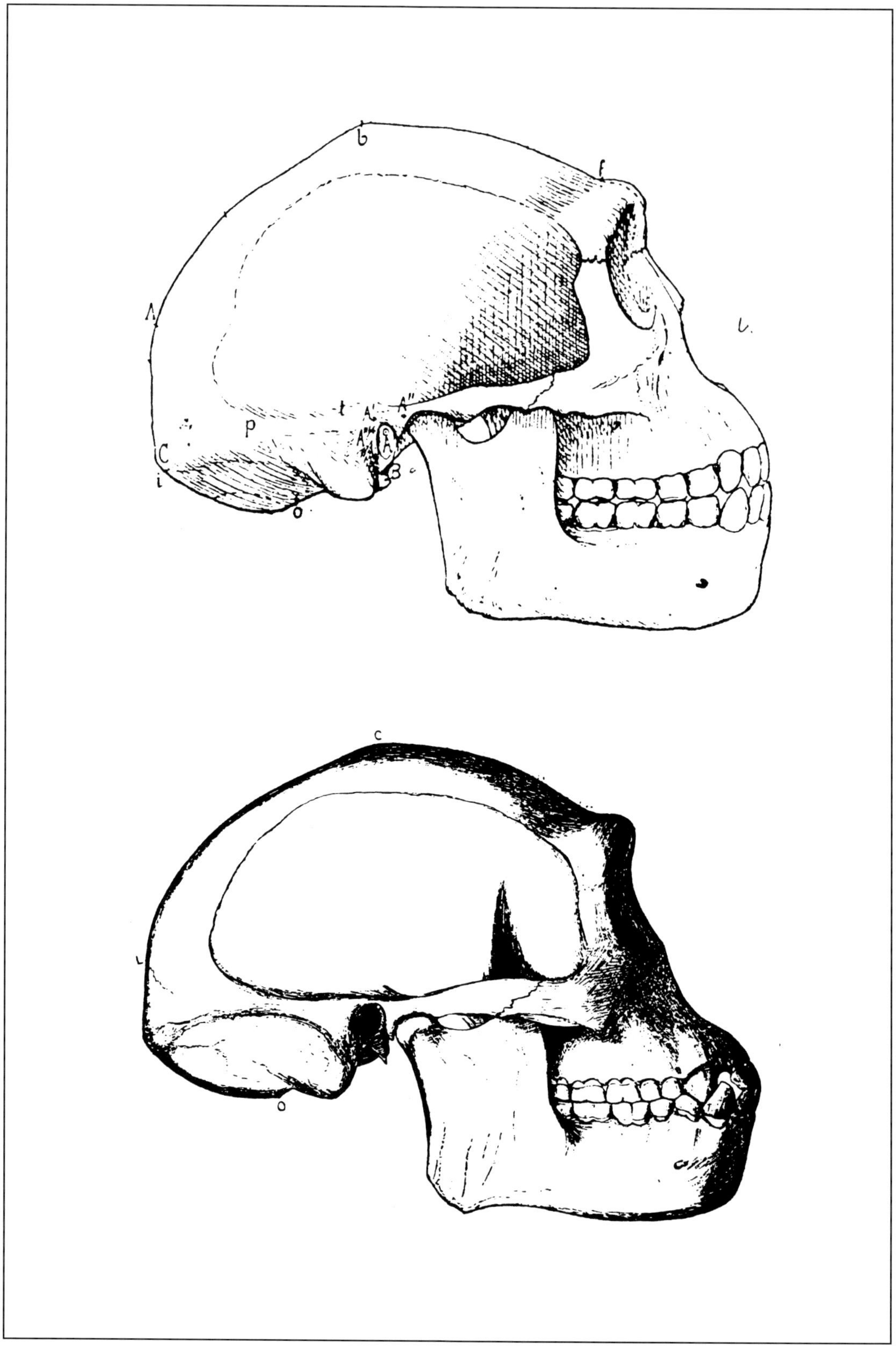

5.14 a + b Comparison of the reconstructions of the Pithecanthropus skull by Manouvrier in 1895 (above) and Dubois in 1896 (below).

5.15 Dubois and other delegates at the Fourth International Congress of Zoology in Cambridge (1898).

5.16 Collection of endocranial casts, made by Dubois for his cephalization studies.

It was not until the second quarter of this century that attention in the search of human origins was diverted to China. Then, in 1926, a team led by the Swedish geologist John Gunnar Anderson discovered two human molars in the quarry on Dragon Bone Hill near the village of Chou Kou Tien, 25 km. southwest of Peking. The lead towards China as a possible location of an early ancestor had stemmed from a fossil molar found as early as 1899 by chance among some "dragon bones" in an apothecary's shop in Peking. It was determined by Max Schlosser of the University of Munich as:"...the third upper left molar of a so far unknown anthropoid ape", so a group of Western anthropologists soon set out in search of the complete skeleton, of what was predicted would be of the First Human in Asia.

Eventually, in 1929 after the anthropologist/anatomist Davidson Black had joined them, a skull was discovered by a Chinese member of the team, the anthropologist Pei Wenzhong. During the ten years' work that lay ahead, unparalleled excavations would produce fossil remains of another 40 individuals, while in a nearby cave about 100,000 stone tools were unearthed. The skull of *Sinanthropus*, 'Peking Man' showed a great similarity, not only to that of 'Heidelberg Man', but also to that of 'Java Man'. It prompted Black and - after his death in 1933 - his successor Franz Weidenreich, a German anthropologist, to assume such a close relationship with Dubois' Pithecanthropus that they considered them all to belong to one group.

Dubois reacted vigorously. Admitting that there was some similarity in the morphology of the skulls, the brain capacity of *Sinanthropus* - in contrast to that of *Pithecanthropus* - was indisputably human. The argument of difference in brain capacity eventually turned against Dubois: in 1934 a Sinanthropus skull was discovered with a capacity as low as about 850 cc. As this fell below the 900 cc. of the Pithecanthropus skull, rendering the latter a hominid, Weidenreich classified it under the same group of Sinanthropus. Unfortunately, the complete collection of fossils unearthed at Chou Kou Tien before the Second World War has disappeared, probably as a result of a Japanese hold-up of their transport by train from Beijing to Tientsin in 1941. What has remained are Weidenreich's accurate description of the fossils and some casts he made of the most important specimen.

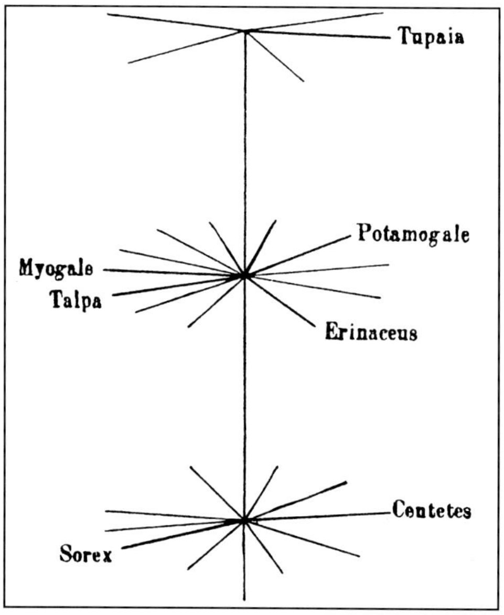

5.17 Dubois' drawing of his phylogenetic 'family-tree' of the mammals showing abrupt leaps in the evolutionary process (1899).

Meanwhile, in 1931 and 1932, a Dutch mining engineer W.F.F. Oppenoorth found eleven skulls in the Solo Valley, near Ngandong. They were named *Homo soloensis*, 'Solo Man'. Dubois, who had compared this new hominid to 'Wadjak Man', soon came to classify Sinanthropus with this, the same proto-Australian race of Solo.

Based partly on his cephalization studies of Pithecanthropus and partly as a reaction to the new discoveries of 'Peking Man' - and its implications for comparison with the 'Java Man' to which now more human-like characteristics were being ascribed - Dubois decided to emphasize the ape-like features in an article *On the Gibbon-like Appearance of Pithecanthropus erectus* (1935).

In spite of the fact that it has never been Dubois' objective to 'lower' the status of Pithecanthropus to that of an Ape, this article created a new myth that he had suddenly revised his thesis completely and now regarded Pithecanthropus as a gibbon-like ape.

Meanwhile, new evidence from Java had emerged in the period between 1936 and 1940, which presented a further threat to Dubois' ideas. The anthropologist Ralph von Koeningswald discovered in Modjokerto and Sangiran, in central East Java, new fossil remains that strongly resembled Pithecanthropus. In 1939, when Von Koeningswald visited Weidenreich and compared his Javanese skulls with the Chinese examples, they were both struck by the similarities and decided to publish a joint article in *Nature* (1939). In this article they strongly opposed Dubois' claim about the gibbonoid features of Pithecanthropus, postulating that both the Javanese and Chinese forms definitively belonged to the human group.

In fact, it was Von Koeningswald and Weidenreich who changed the name of *Pithecanthropus erectus* to *Homo erectus*, which has remained the official name.

The term 'pithecanthropine' is now increasingly used only to indicate the Asian or Far Eastern examples of *Homo erectus* which, although most probably derived from African ancestry, bear some features that in the opinion of some experts may justify their being considered distinctive (*cf.* Leakey 1989).

Dubois replied in strong terms, largely rejecting Von Koeningswald's discovery

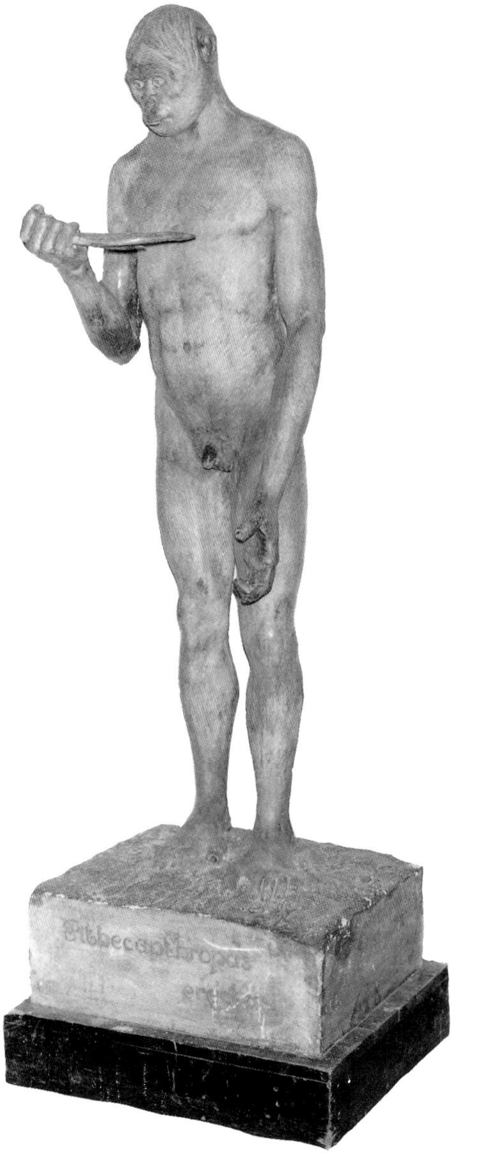

5.18 Dubois' 'life-size' reconstruction of Pithecanthropus for the World Exhibition in Paris in 1900.

5.19 *Excavations at Trinil by the Selenka Expedition in the period from April - August 1907.*

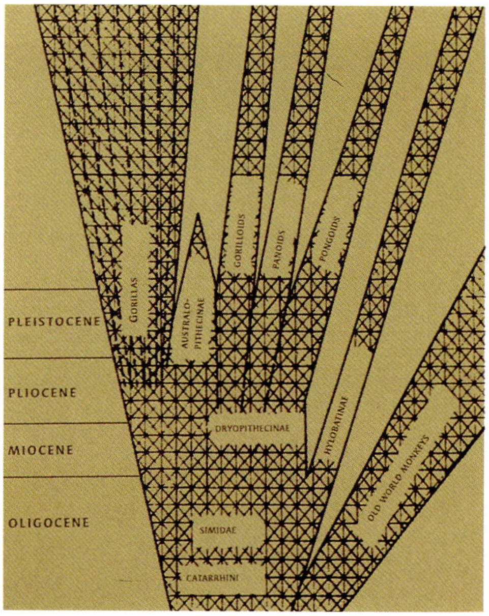

5.20 *Franz Weidenreich's phylogeny of the primates (1946).*

5.21 *The casts of the Chou Kou Tien fossils from China, made by Franz Weidenreich, shortly before the real specimen disappeared in 1941.*

of a 'new' Pithecanthropus, ascribing it to *Homo soloensis*. Shortly before his death in 1940, he decided to bundle his reactions in a trilogy, and published them in the *Proceedings* of the Academy. While he reassessed the new discoveries from South-East Asia, referring to them all as belonging to the proto-Australians, he claimed an exception for Pithecanthropus from Trinil, which he firmly believed to be the only real 'Missing Link'.

However, largely due to the incomplete fossils from Java and the debate on human origins then still in its infancy, a generally accepted and conclusive judgement on the exact status of Dubois 'Ape-Man' could not yet be passed.

Ironically, the final answer to the place of Pithecanthropus in the human 'family-tree' was not to be found on Java, but would eventually be formulated elsewhere.

Just at the time that the interest into the study of human ancestry and its place in nature had increased even more, and others prepared to set out in search for human fossils, Dubois himself withdrew from the debate and locked his fossils away. This curious behaviour soon gave rise to various rumours and speculations about his attitude and state of mind.

As John Reader (1988) notes, Dubois grew rather bored with the scientists who came to

5.22 Eugène Dubois meets Franz Weidenreich (left), L.D. Brongersma (second from right) and H. Boschma (right) at the National Museum of Natural History in Leiden (1935).

5.23 Photograph of Eugène Dubois amidst his family and colleagues in the Senate Room of the University of Amsterdam on the occasion of his retirement in 1928.

visit him and, after examining the fossils from Java, opposed his views. His correspondence from that time confirms his irritation with scientists who did not share his opinion. As he grew older, his almost paranoid suspicion even led him to believe that 'the Catholics' were out to destroy his fossils. One day, the rumour was spread that Dubois himself had destroyed the fossils and abandoned his belief in evolution. In expiaton for his sins and as a pious deed towards his sister, a nun, he was said to have returned to the Roman Catholic faith.

Bert Theunissen (1989) introduces another explanation for his peculiar attitude. In view of Dubois' humble professorship at the University of Amsterdam, and his ambition to 'exploit' his fossil vertebrate collection to further his own career, he decided to keep them away from others, and first produce his own, revised description. However, his full treatment of the vertebrate fossils from Java failed to materialize, while outsiders were denied access to study them properly. Finally, after two decades, the *Koninklijke Nederlandsche Academie voor Wetenschappen* (Royal Netherlands Academy of Arts and Sciences) intervened and Dubois agreed to lift the embargo.

Meanwhile, Dubois, who clearly preferred fieldwork to the laboratory, had actually started to devote more time to his studies on the biology, palaeontology, and ecology of Limburg. Now living on a large estate in the countryside of southern Holland, he became engaged in several ecological studies and experiments that would soon add a new dimension to his perspective on the place of man within nature.

6.1 Satellite photograph of Lake Turkana including Koobi Fora at the protruding point on the lower east side.

6 AFRICA: THE 'CRADLE OF MANKIND'

Although Dubois himself now virtually withdrew from the debate on Pithecanthropus he himself had initiated, the first quarter of this century witnessed an increased interest in human ancestry. Hundreds of articles appeared on Pithecanthropus, Neanderthal, Cro-Magnon, and the forged 'Piltdown Man', while many popular-scientific works focused their attention on the origins of mankind. The romantic image of nineteenth-century 'Primitive Man' detached from any apish ancestry gradually gave way to artistic interpretations and reconstructions of prehistoric man, that were based on accumulated fossil finds.

At first, the only significant fossil evidence retrieved since the heated debate on Pithecanthropus flared at the beginning of the twentieth century came initially from rather incidental finds in Western Europe. On 21 October, 1907, a human jaw was found in a sand-pit near Mauer in Germany. After investigation, Otto Schoetensack initially ascribed it to what he created into a new species: *Homo heidelbergensis*. Based on ecological evidence of the site, its age was estimated at not less than 500.000 years old! However, it was important that, using fossil evidence, it was now assumed that human ancestors lived in Europe in a very early period of time. In the same way, in 1908, a skeleton was found at La Chapelle-aux-Saints in France, described by Boule the same year as that of another Neanderthal.

6.2 Bust of Pithecanthropus by the sculptor Viktor Haberl Jr. in collaboration with the palaeontologist Maria Motl (1930).

In the course of the twentieth century, more pithecanthropine skulls have been found in Indonesia by Von Koeningswald, the *Nederlandsch-Indische Geologische Dienst* (Netherlands-Indies Geological Service), and its successor the Indonesian Geological Survey. As Teuku Jacob (1979) documents, most human fossils have been found in East Java, rendering Sangiran to one of the oldest and richest find-spots in Asia.

6.3 Zdeněk Burian's painting of 'Neanderthal Man' based on the evolutionary reconstruction of fossil evidence from Europe (around 1950).

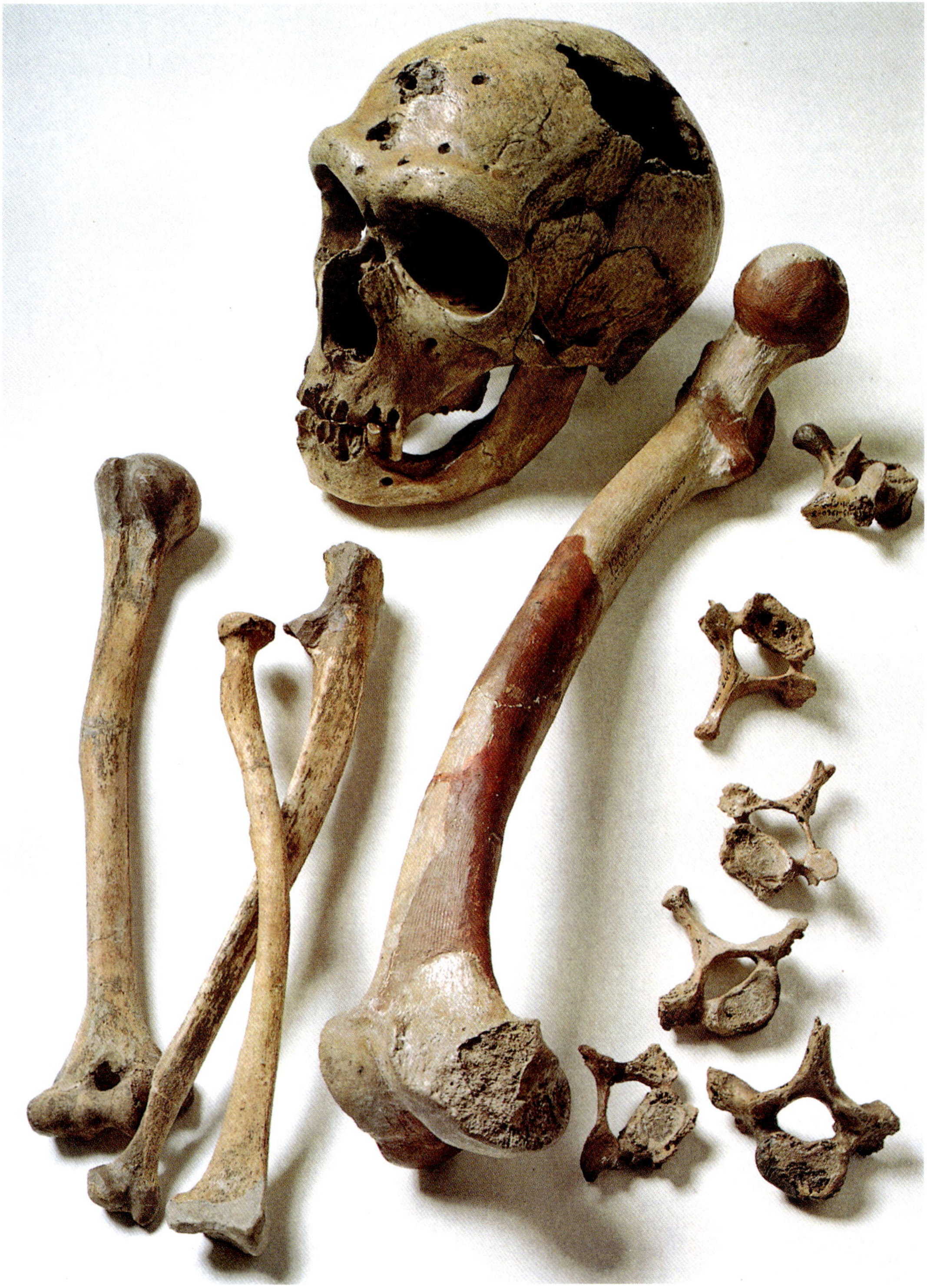

6.4 *Fossil remains of the 'Old Man' found at La-Chapelle-aux-Saints. Deformation of several bones as a result of arthritis led scientists wrongly to believe that Neanderthals were directly related to the apes, not to modern humans.*

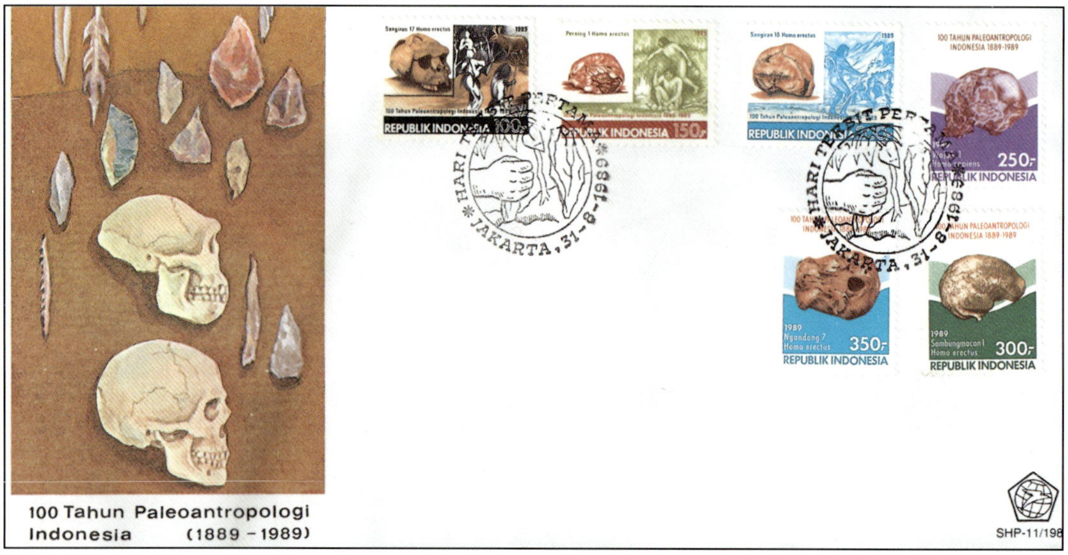

6.5 In 1989, Indonesia celebrated its First Centennial of Palaeoanthropology by a special issue of postage stamps.

6.6 Skull of an Orang-utan, elaborately incised by the Kenyah people of Kalimantan.

Primarily preoccupied with the fossil finds of Neanderthals and Pithecanthropi in Europe and Asia, the attention of the scientific world had so far virtually ignored the potential role of the African Continent in human origins.

Nevertheless, for a considerable time Raymond Dart and Robert Broom had been engaged in the pioneering search for early man in Southern Africa. With the help of his students, Dart had discovered the fossils of an ape-like creature in a limestone quarry near Taung in Transvaal in 1924. The small skull had a remarkably larger capacity than that of a baboon, while its jaw was certainly smaller. He noticed it was more 'primitive' than 'Java Man' or 'Peking Man', and finally named it *Australopithecus africanus* (Southern Ape from Africa).

His announcement of the 'Taung Child' in February 1925 was greeted with scepticism, but after some time massive

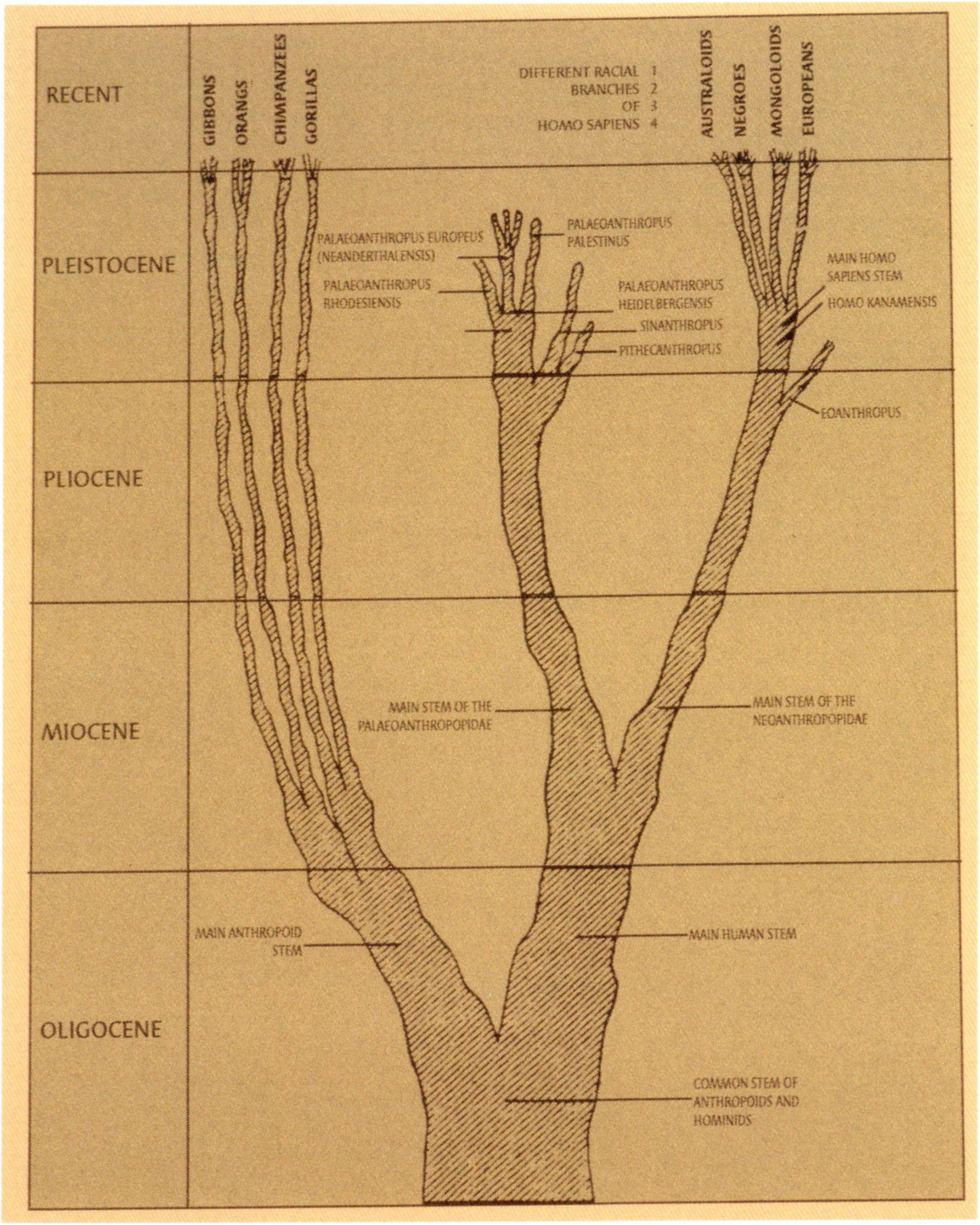

6.7 *Phylogeny of primates by Louis Leakey (1924).*

fossil evidence proved him right. At five locations in South Africa, hundreds of fossil fragments of Australopiticinae, which eventually could be divided into two upright-walking species: *robustus* and *fragile,* were unearthed.

Broom's estimation of the age of *Australopithecus* as two million years old was criticized even more as only a few people could accept the hypothesis that a distant human ancestor was probably an upright-walking creature which had lived in South Africa and which had a brain capacity comparable to that of a chimpanzee!

The 'robust' type of Australopithecus was characterized by a heavy jaw and larger molars; the 'fragile' by smaller molars and shorter stature. The dilemma posed by such classification was that the more 'primitive' *robustus* dated from about one million year later than the *fragile,* the jaw and molars of which appeared more human-like. Had *africanus* been the ancestor of *robustus* ?

It was only long after the Second World War that *Australopithecus* was eventually accepted and new 'pieces' were brought in from elsewhere in Africa to fit into the complex jigjaw puzzle of human ancestry. Indeed, the solution came many years later from the north, from the pioneering work of the anthropologists Louis and Mary Leakey. In the course of almost three decades, through their persistent efforts and unflagging energy, they mapped out the history of the Olduvai Gorge in Tanzania, and with this, the early history of mankind in Africa.

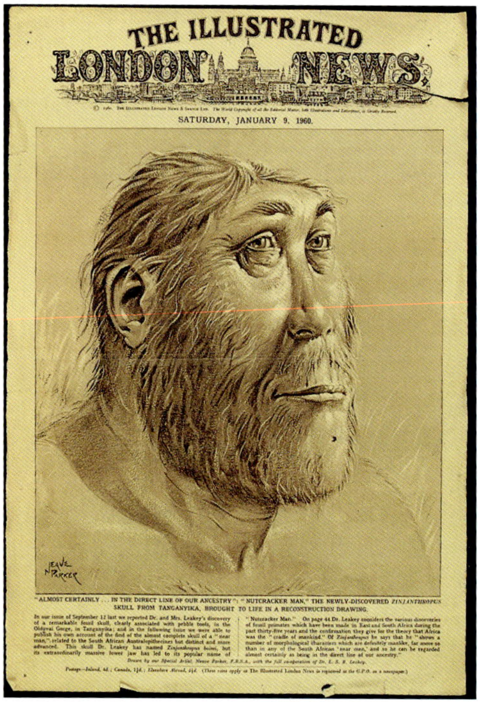
THE ILLUSTRATED LONDON NEWS

SATURDAY, JANUARY 9. 1960.

"ALMOST CERTAINLY . . . IN THE DIRECT LINE OF OUR ANCESTRY": "NUTCRACKER MAN," THE NEWLY-DISCOVERED *ZINJANTHROPUS* SKULL FROM TANGANYIKA, BROUGHT TO LIFE IN A RECONSTRUCTION DRAWING.

In our issue of September 12 last we reported Dr. and Mrs. Leakey's discovery of a remarkable fossil skull, clearly associated with pebble tools, in the Olduvai Gorge, in Tanganyika; and in the following issue we were able to publish his own account of the find of the almost complete skull of a "near man," related to the South African Australopithecines but distinct and more advanced. This skull Dr. Leakey has named *Zinjanthropus boisei,* but its extraordinarily massive lower jaw has led to its popular name of "Nutcracker Man." On page 44 Dr. Leakey considers the various discoveries of fossil primates which have been made in East and South Africa during the past thirty-five years and the confirmation they give for the theory that Africa was the "cradle of mankind." Of *Zinjanthropus* he says that he "shows a number of morphological characters which are definitely manlike, far more so than in any of the South African 'near men,' and so he can be regarded almost certainly as being in the direct line of our ancestry."

Drawn by our Special Artist, Neave Parker, F.R.S.A., with the full co-operation of Dr. L. S. B. Leakey.

6.8 The discovery of the Zinjanthropus skull from Tanzania by Mary and Louis Leakey in 1959 made frontpage news. The Illustrated London News of 9 January, 1960, published the 'Nutcracker Man' 'brought to life in a reconstruction drawing'.

In 1959 they discovered the important 1.75 million year old skull of the robust *Zinjanthropus boisei,* 'Dear Boy', the *first* early human fossil from East Africa, later referred to as *Australopithecus bosei.* Soon afterwards, in 1960, the Leakeys discovered the remains of a more refined, human-like creature - the first fragments of which their young son Jonathan found - to which the use of a whole range of unearthed artefacts could be ascribed: the new species of *Homo habilis,* the 'Handy Man'. One of the particular characteristics of this species of the *Homo* line was its use of the Acheulian handaxe, appearing in 1.4 million-year-old deposits at Olduvai. An article by Louis Leakey, Phillip Tobias, and John Napier in *Nature* (1964) described the discovery of this new species of hominid, the ultimate ancestor of modern humans. These and ad-

6.9 *Bayard's romantic image of 'Primitive Man' was far removed from the theory of evolutionary descent from an apish ancestor (1870).*

ditional finds from Olduvai led then to the extension of the trunk of the early human 'family-tree' to include three different types of Ape-Men in South and East Africa. All belonging to the family of *Hominidae*, and the genus *Australopithecus*, they represent the species *africanus, robustus,* and *bosei.*

In this way the descent of man on the African continent had suddenly been traced back about two million years, a point in time at which the species *Homo* had emerged in East Africa and had started to use tools. However, the problem of exact dating and the associated issue of the relationship with *Australopithecus africanus* still remained. In order to gather more data to substantiate the fossil evidence for the various hypotheses on the relationship between the different hominid types by this time retrieved in Sub-Saharan Africa, a large expedition was mounted in 1967 to search for fossils in the area of the Omo Valley in Southern Ethiopia. The Omo expedition included the French team of Camille Arambourg, the American team of Clark Howell, and the Kenyan team of Richard Leakey. Although the Omo discoveries themselves were rich but not spectacular, its narrative would unexpectedly lead to one of the most important decisions in the history of anthropology.

The French and Americans retrieved many fossil remains of a whole range of species of antelopes and pigs, completed with fragments of skulls, bones, and teeth of both *Australopithecus bosei* and *Homo habilis* at their excavation sites. Due to the absence of early fossils in their area, the Leakey team decided to discontinue their work. Only fragments of two relatively recent human skulls, about 100,000 years old, were found. On the way back to Nairobi, a huge thunderstorm forced the co-author of this book Richard E. Leakey to make a detour around the eastern shore of Lake Turkana. Then, from the air, he closely observed the stratified sediments of the shore clearly revealing their fossil potential and, a few days later, a reconaissance flight firmly established his expectations.

6.10 Turkana doll, ornamented with beads and shells by the Turkana people, North-West Kenya.

A few months later, at the Headquarters of the National Geographic Society in Washington D.C., after reporting on the Kenyan part of the Omo Project - instead of requesting additional funding for the Omo research - new, unexpected plans were proposed for an independent Leakey expedition to Koobi Fora on the eastern shore of Lake Turkana. The Board agreed.

The lake, which had been able to hide its treasures throughout the ages and postpone its own "discovery" as the last in the chain of Great African Lakes until the end of the nineteenth century, was about to yield up its secrets. Its inhospitable, volcanic environment and difficult access had protected Lake Bussa, as it is locally known, against invasion. When the Hungarian Count Samuel Teleki and his party finally reached the area on 4 March, 1888, he wrote in his diary that they: "...were rewarded for all our frightful hardships, when we saw the long sought-for lake, shimmering before us in the distance." He named it Lake Rudolf after Crown Prince Rudolf of Austria. In 1963, after Kenyan Independence, it was renamed after the Turkana people who live on its shores: Lake Turkana.

The preliminary exploration of the lake shore in 1968, in which the Kenyan fossil prospector Kamoya Kimeu also participated, firmly established the potential for successful work at this new site for years to come. Indeed, since that first excursion to Lake Turkana, the Leakey team, which includes a large number of researchers, has unearthed more than 200 hominid fossils. In the summer of 1969, a very unusual kind of discovery was made during an expedition on camels to extend the survey right up to the border with Ethiopia. In a dry river-bed an intact cranium of a hominid was discovered. It was remarkably similar to 'Zinj', the *Australopithecus bosei* skull found by Mary Leakey at Olduvai in July 1959, almost ten years to the day of its discovery. After establishing a permanent lakeside camp at Koobi Fora, more and more fossils were discovered in this unique area.

In 1972, Leakey's team, in which the Kenyan 'fossil hunter' Bernard Ngeneo had now also come to work, discovered the cranium that has radically changed our previous ideas about our evolutionary history, the famous '1470'. After reconstruction of the hundreds of pieces by Meave Leakey, and later by the anatomist Alan Walker, it became clear that the cranium was of the same type that Louis had named *Homo habilis,* albeit a little older. It was the earliest evidence that the species *Homo* existed at Koobi Fora, and as such, its discovery received tremendous publicity.

There was another aspect to the discovery of the '1470' skull. Richard Leakey returned to Nairobi, anxious to show his father - who was about to leave for England - the new find. Louis was delighted to touch the final proof of the ideas he had held throughout his career. His hypothesis that the *Homo* line was much older than had so far been supposed was finally confirmed by a discovery of his son. Unfortunately, shortly after arriving in London, Louis Leakey died.

As significant traces of two basic forms or species of human ancestors - *Australopithecus* and *Homo* - similar in bipedalism but distinct in food consumption had now been found at Lake Turkana, an unusual pattern of human evolutionary history was unravelled: *Australopithecus* and *Homo* had lived here at the same time, and probably even in the same area. As Richard Leakey assumed " the Y-shaped pattern seemed to be present: *Australopithecus* on one hand, *Homo* on the other."

This pattern became even more fascinating in 1975 when a remarkable specimen of a cranium was found in the Koobi Fora region. The KNM-ER 3733, retrieved from below the level of the 1.6 million years old Koobi Fora Tuff, proved to be the most complete *Homo erectus* skull so far discovered in Turkana. Until about a decade later the increasing number of fossils of *Homo erectus* remained limited to fragments of the skull, and a few pieces of the rest of the skeleton. Comparable to handaxes from the Olduvai Gorge, similar artefacts, though distinctly older, have also been found in association with *Homo erectus.*

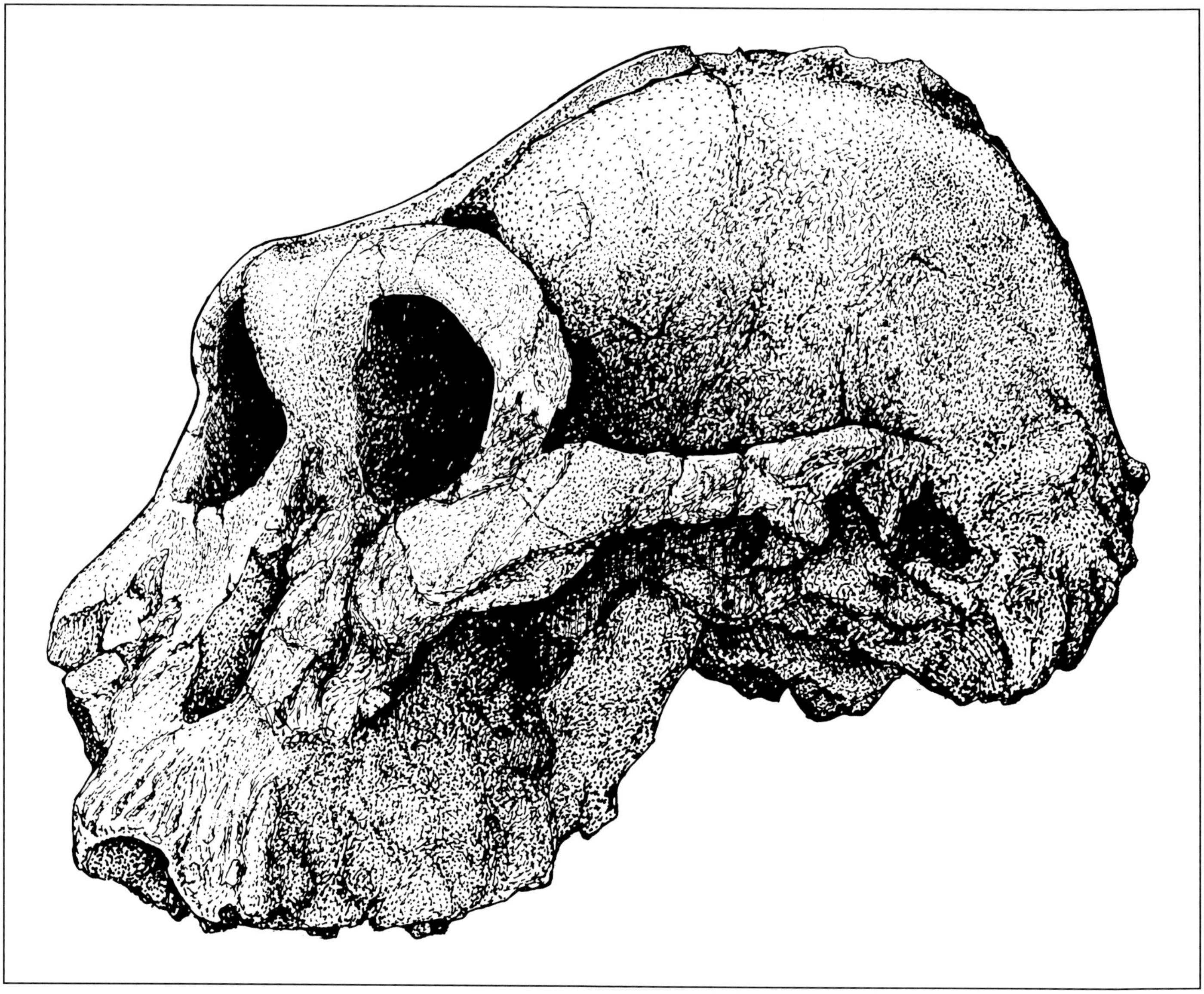

6.11 'Life size' drawing of a complete Australopithecus boisei skull found by Richard Leakey near Ileret on the eastern side of Lake Turkana in 1969. This specimen (KNM ER 406) is 1.6 million years old.

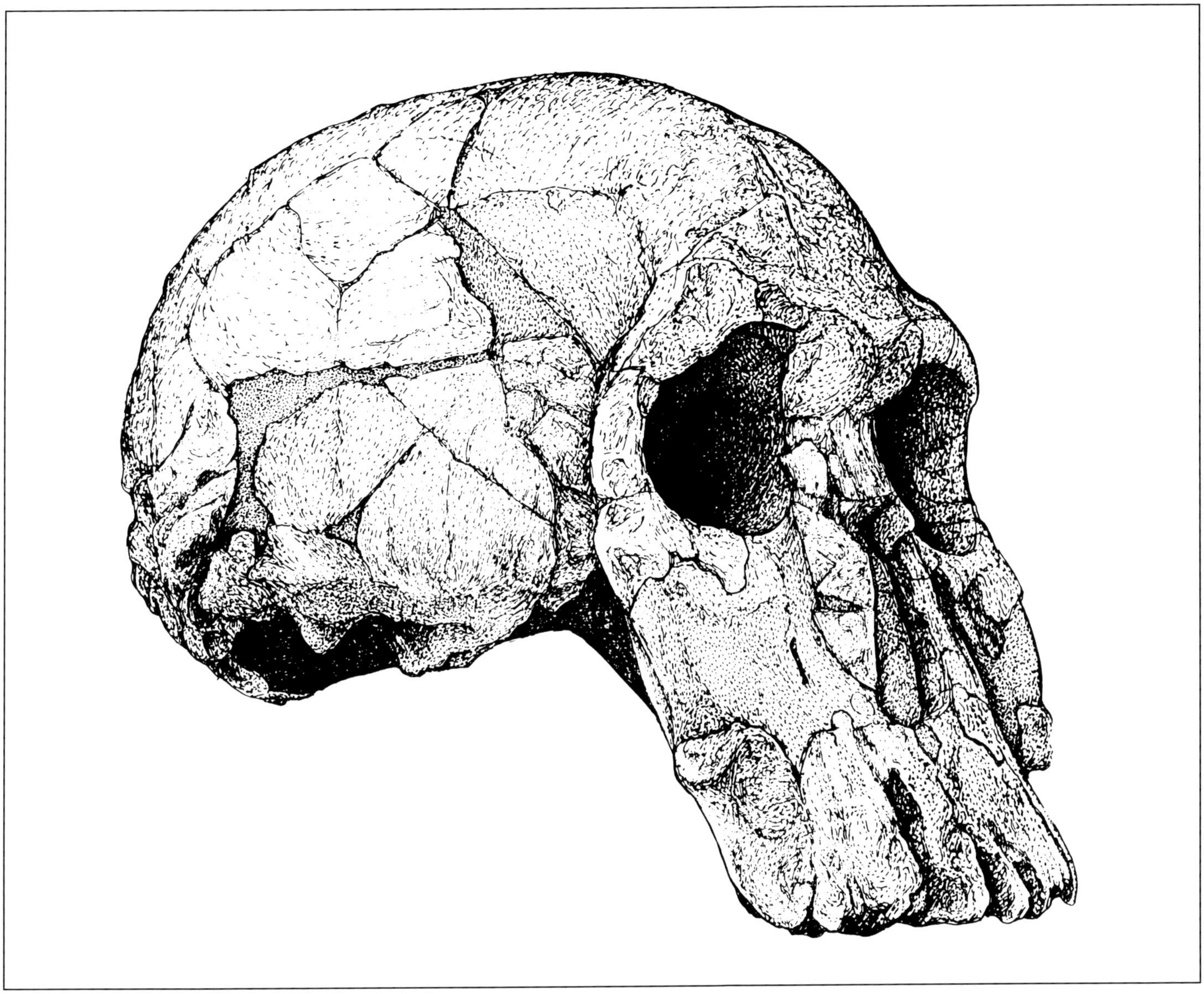

6.12 'Life size' drawing of Homo habilis skull found by Bernard Ngeneo at Koobi Fora on the eastern side of Lake Turkana in 1972. This specimen (KNM ER 1470) is 2 million years old.

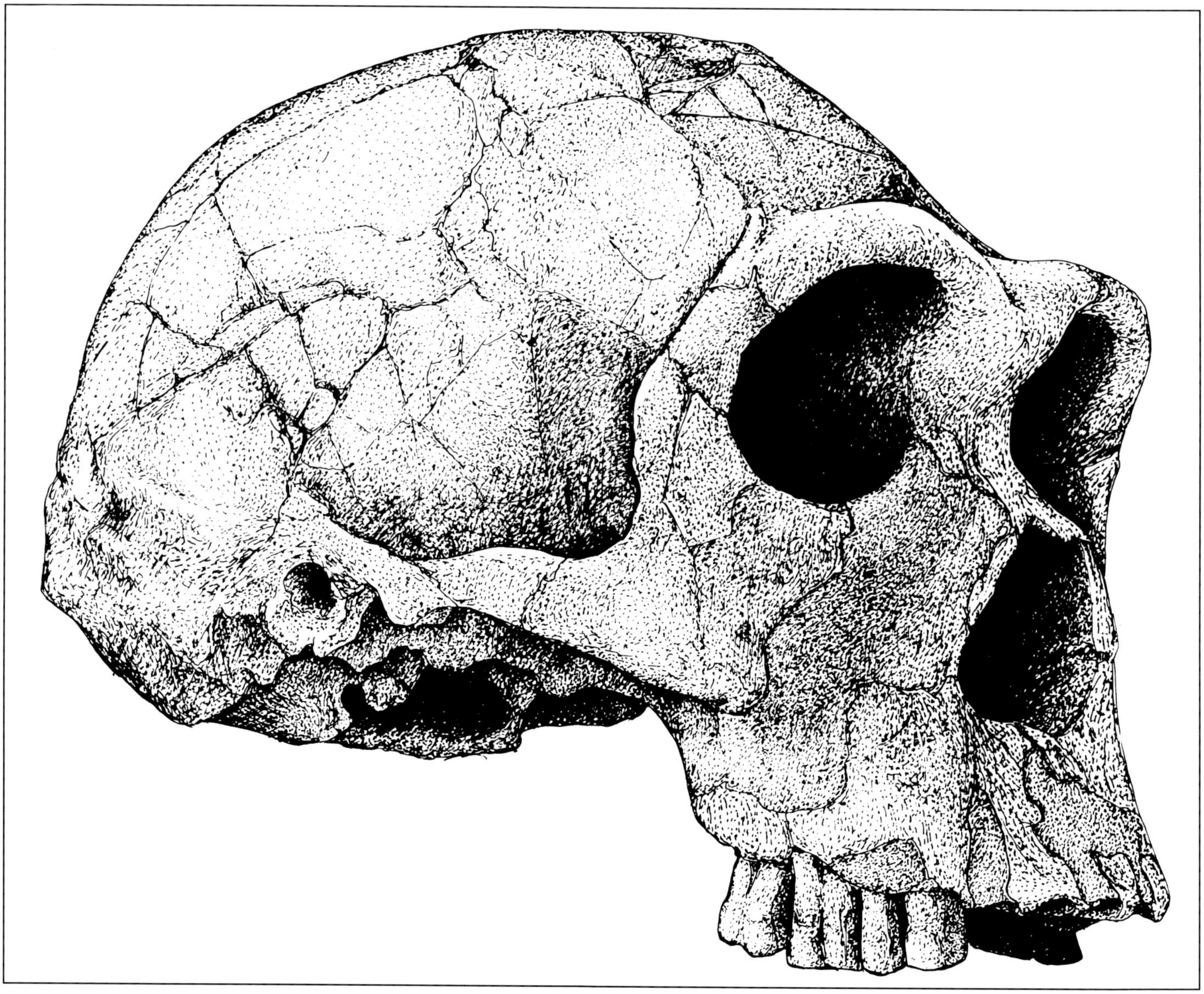

6.13 *'Life size' drawing of Homo erectus skull found by Bernard Ngeneo at Koobi Fora on the eastern side of Lake Turkana in 1975. This specimen (KNM ER 3733) is 1.7 million years old.*

Around the same time, in the mid-1970s, a most remarkable discovery was made at Laetoli, 20 miles south-west of the Olduvai Gorge. Mary Leakey found a set of hominid footprints close to 3.6 million years old. Clearly, three hominids had walked north leaving their footprints in the fresh volcanic ash from the nearby volcano Mount Sadiman. The Laetoli footprints show us that, as early as more than 3.5 million years ago, hominids walked upright as modern humans do, leaving their hands free for carrying artefacts, food, and other items.

At about the same time, several fragments of jaws and teeth were found at the same site, which in *Nature* in 1976 were ascribed to a hominid, *i.e.* to *Homo*. Since these fossils were 3.6 millions years old, the *Australopithecus-Homo* pattern could be extended as far back as 3.6 millions years ago.

6.14 Dr. Mary Leakey measures the famous footprints she discovered in Laetoli, Tanzania (1978).

Almost at the same time as the publication of the Laetoli finds, Don Johanson and Tim White announced the results of their excavations in Ethiopia since 1973. The fossil remains included parts of a skeleton of a creature later known as 'Lucy' and of about fifteen different individuals, the so-called First Family. Given the range of size of the fossils, and the extensive range of anatomical variability, initially three different species were identified: two *Australopithicinae*, one large and one small, and an early *Homo*.

Later on, Johanson and White changed their assessment and concluded that the fossils from Hadar represented just one, albeit variable, species instead of three: the *Australopithecus afarensis*, the trunk of their 'human bush' from which all hominid species would have evolved. A less likely scheme. This reassessment of the Ethiopian fossils led to another supposition that the Laetoli fossils would be from the same species as those from Hadar. As this yet another heated debate in palaeoanthropology raged, described by Roger Lewin in his 'Bones of Contention' (1987), initially most anthropologists tended to support the *afarensis* hypothesis. But a few exceptions remained (*cf.* Leakey and Lewin 1992). Recently, in 1991, Johanson's Institute of Human Origins announced that new fossil finds in Ethiopia in 1990: "..may rekindle arguments that A. afarensis actually consists of more than one species."

Meanwhile fieldwork near Koobi Fora had progressed well, and in August 1984, at Kamoya's *Homo erectus* site near the sandy bed of the Nariokotome River, a piece of human skull no larger than a matchbox was collected from among black lava pebbles. As the excavated soil was thoroughly sieved later on, the word was soon spread: "*Tumepata Kishwa!- We have found the skull!*" Gradually, other parts were found, and over the next three weeks, almost an entire *Homo erectus* skeleton of a boy of about 13 years old was unearthed. A specta-

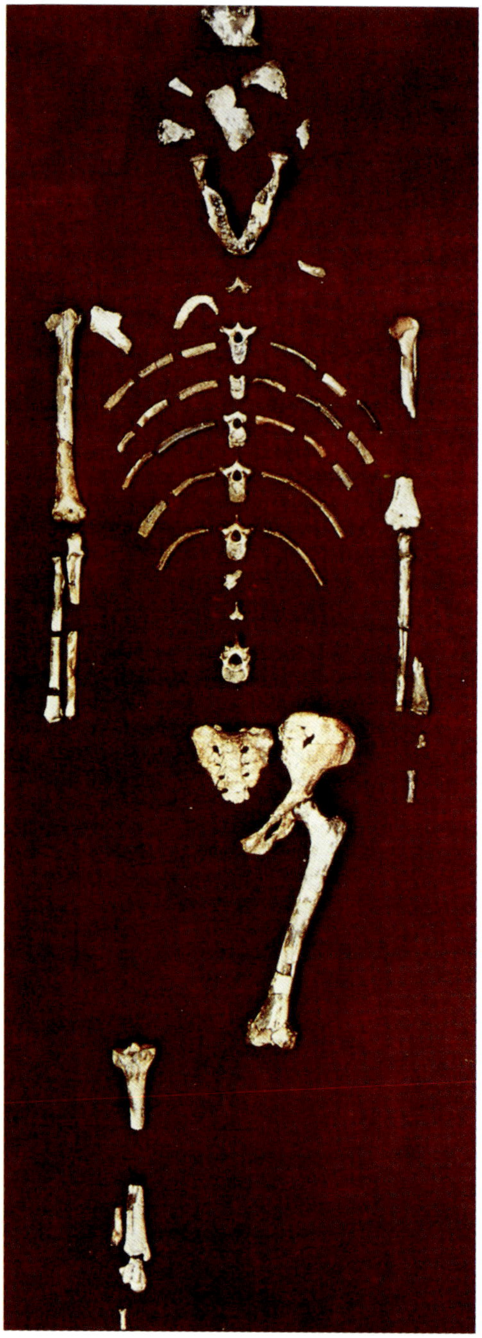

6.15 Fossil remains of almost half of the skeleton of 'Lucy', Australopithecus afarensis, discovered by Don Johanson in Hadar, Ethiopia in 1974. Its estimated age is about 3.5 million years old.

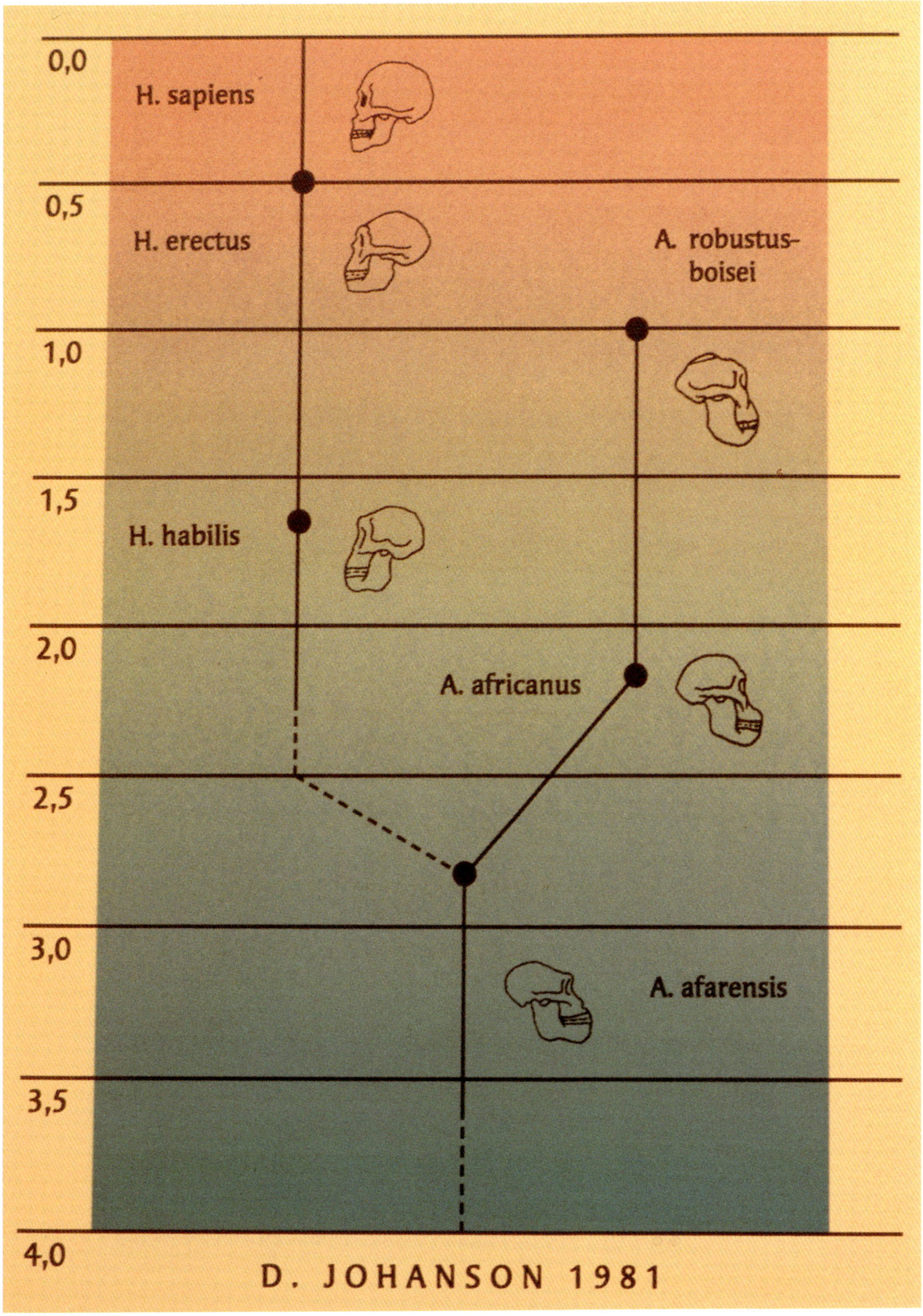

6.16 *'Family-tree' of human ancestry by Don Johanson (1981).*

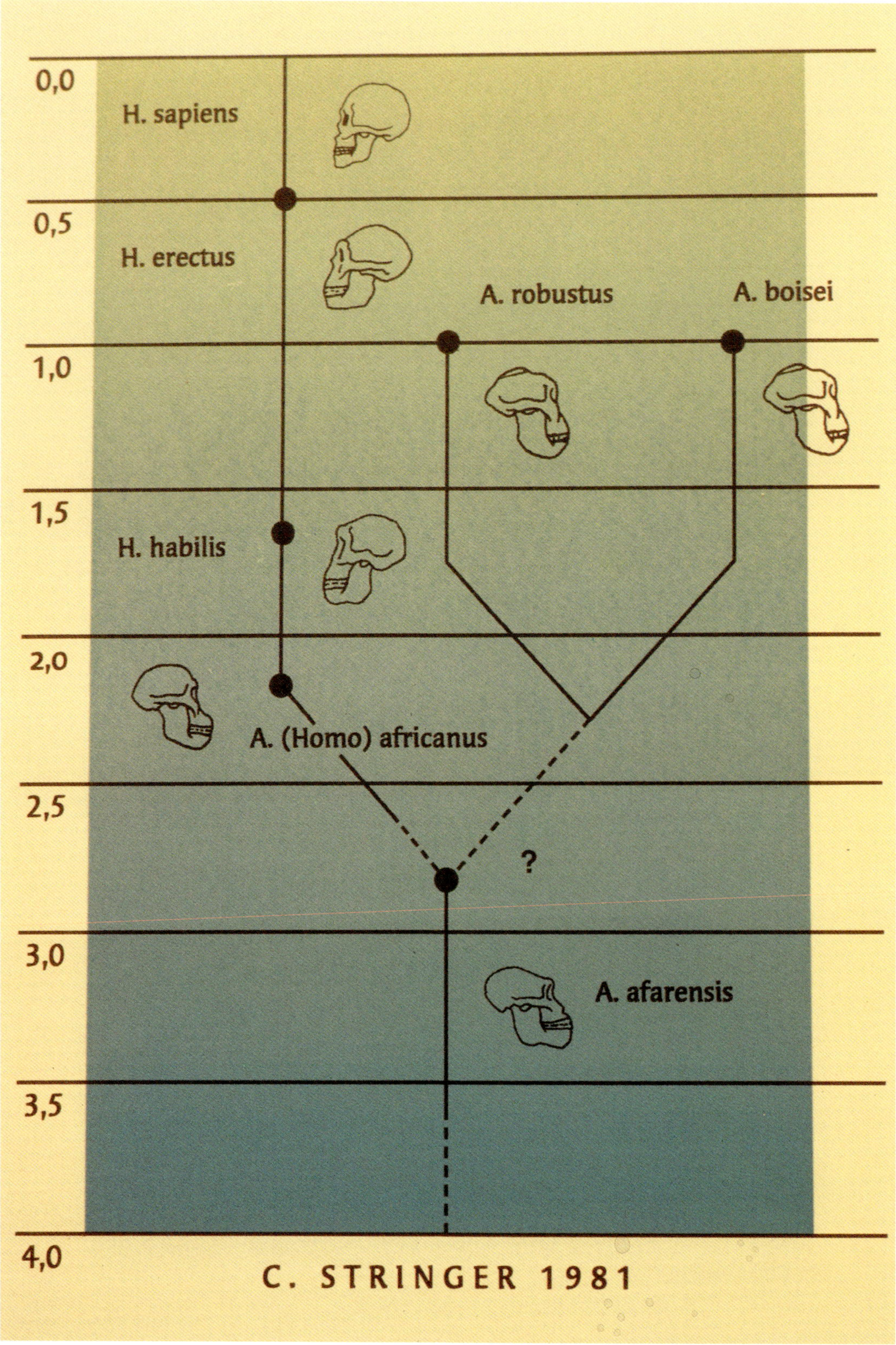

6.17 'Family-tree' of human ancestry by Chris Stringer (1981).

cular discovery of the earliest known set of one individual's bones retrieved *in situ*. Moreover, it was of the extraordinary age of 1.6 million years old.

As Mary Leakey remarked while visiting the excavation site: "You have to go to Europe, to Neanderthal graves, to see fossil skeletons as complete as this." Indeed, to find such a complete skeleton, we must jump forward to 100,000 years ago when Neanderthals in Europe started to bury their dead.

The significant discovery of the 'Turkana Boy', who as a representative of our ancestors lived on the African savannah, has also provided us with a unique view on the growth and development of early humans. It clearly opposes the idea that over the millennia humans have gradually grown larger. From the 'life-size 'of the boy from Turkana, we now know that the general size of present humans had already been acquired some one and a half million years ago.

With these important discoveries in Africa, the reconstruction of the '*Human Bush*' (Richard Leakey 1992), could now be further filled in. The trunk of the tree, representing one single species, bushes out with the two main branches of the *Australopithecus* and *Homo* species. Since one branch broke off as *Australopithecus* became extinct, only one branch seems to have continued: *Homo sapiens.*

Later discoveries and analyses of hominid fossils such as the 3735, a 1.9 million year old partial skeleton unearthed in 1975, would establish, that there have been three separate species in the immediate 'pre-*erectus* period': *Australopithecus robustus; Homo habilis;* and a third species with ape-like proportions in its arms and legs. This pattern may very well

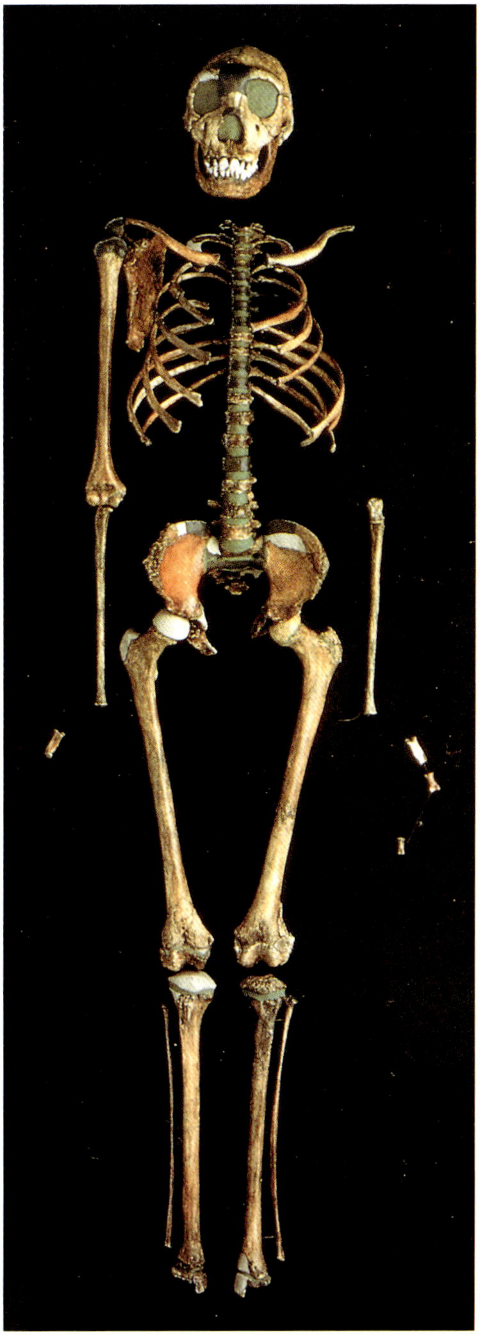

6.18 *Skeleton of a 1.65-million-year-old Homo erectus boy discovered by Kamoya Kimeu from the Leakey team at Nariokotome to the west of Lake Turkana in 1984. The'Turkana Boy' fossils are the most complete remains of an early human ever found.*

6.19 Major excavation sites of hominids in Africa and Europe.

extend back at least three million years, probably even further (*cf.* Leakey and Lewin 1992). Future finds and views could give the trunk of this 'Human Bush' a more complex pattern that might perhaps evolve beyond the Y-shape of the basic two branches.

Eventually it had taken almost one century and another continent to find the first, essentially complete skeleton of a *Homo erectus* individual since the species was discovered by Dubois on Java.

Before Dubois, Charles Darwin and Thomas Huxley had recognized the anatomical similarities between humans and apes, notably African apes, such as the chimpanzee and gorilla. This led them to the assumption that Africa was the 'Cradle of Mankind'. This thesis, based on comparative anatomical studies between man and ape, has now been sufficiently confirmed by Africa's fossil record.

Recent molecular biological studies of proteins and genes in living creatures have been able to reconstruct 'family-trees' based on the 'molecular clock'. Pioneers in modern molecular anthropology such as Morris Goodman (1986) have also confirmed Darwin and Huxley's interpretations that humans and African apes were closely related, separating both from the Asian great apes.

The fossil record has also established that all previous hominid species older than approximately one million years existed exclusively in Africa, specifically in South and East Africa.

Dubois discovered the fossil remains of his *Pithecanthropus* - a member of the *Homo erectus* group - while, almost one million years ago, it was 'on the way' outwards from Africa into South-East Asia and Oceania. It is understandable, of course, that Dubois, who had followed Haeckel in his assumption that human origins were located in *Lemuria,* supported Asia as the birthplace of humankind. His opinion was naturally strengthened by his own discovery of the 'Missing Link' in Java.

Since Dubois' time, however, the reconstruction of the origins and spread of *Homo erectus,* largely based on his new palaeontological approach, has been traced back from the former Netherlands East Indies via China and South Africa to establish East Africa as the 'Cradle of Mankind' by those who followed in his footsteps.

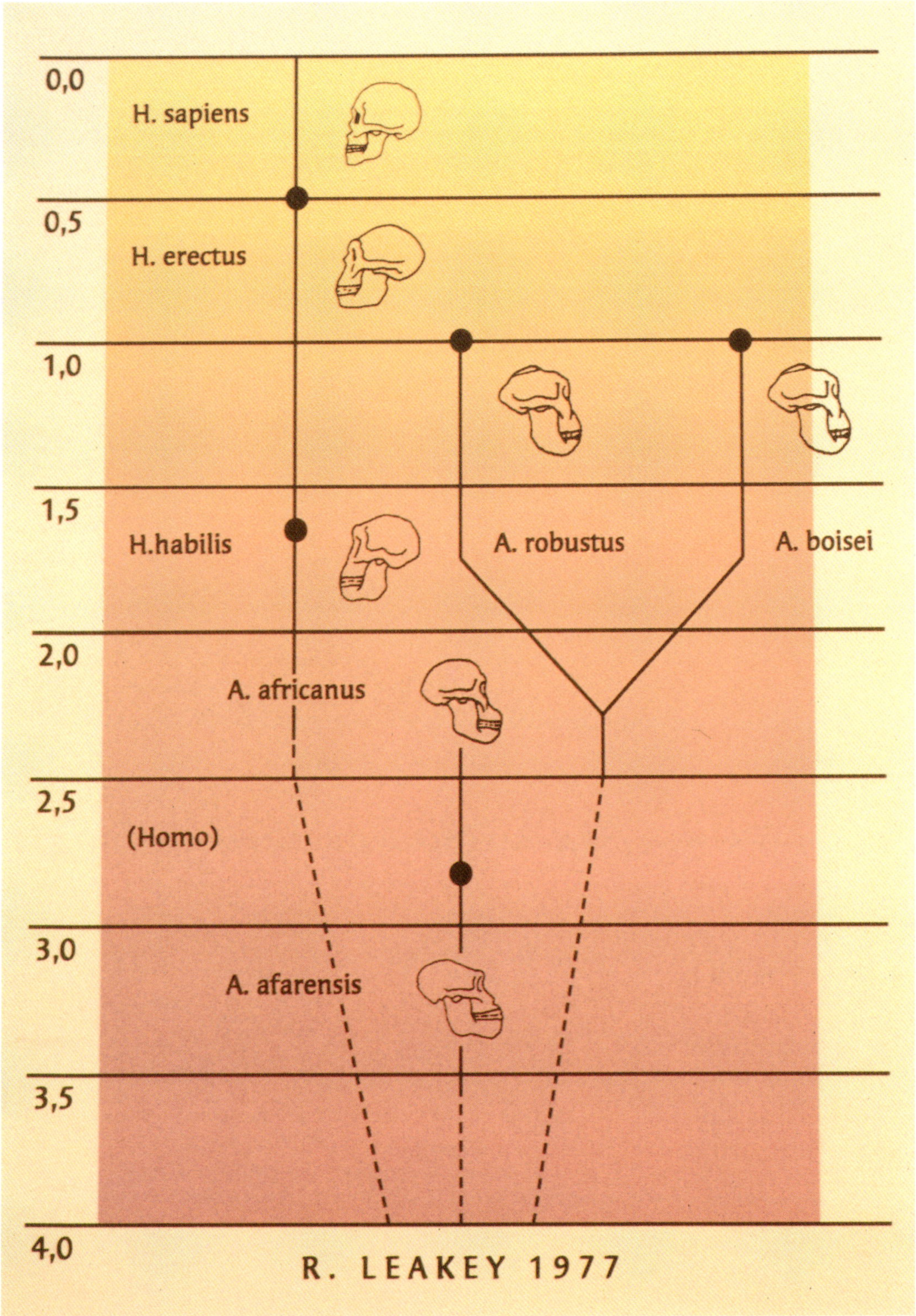

6.20 'Family-tree' of human ancestry by Richard E. Leakey (1977).

Indeed, as we now know, the significant turning point in human history denoted by *Homo erectus* was its capability to mount the first hominid migration from Africa onto the other continents. As nomadic people, *Homo erectus* groupings travelled to new lands, not so much driven by migratory pressure but more by an unconscious drift, into Europe and Asia. Although the colonization of new continents substantiates the new capabilities of *Homo erectus,* probably most human ancestors remained in Africa. It is remarkable that, in contrast to the African context of *Homo erectus,* stone tools such as the Acheulian handaxes were not characteristic among stone-tool industries in Asia.

The reconstruction of the first human migration out of Africa also has implications for an issue in palaeoanthropology that has become a topic in recent debate: the origin of modern humans. The period of time covered since the days of the 'Turkana Boy', 1.6 million years ago, up until about half a million years ago has shown the evolution and disappearance of *Homo erectus.* Modern humans, *Homo sapiens sapiens* are now generally accepted to have first

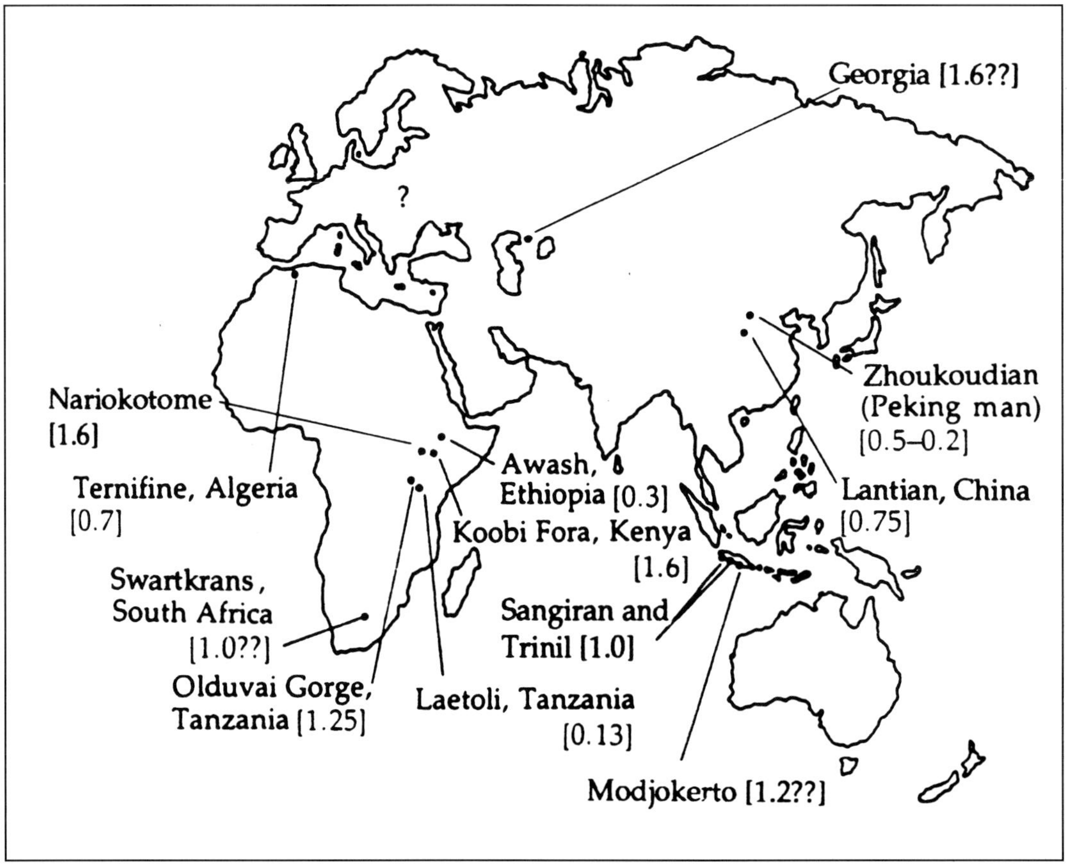

6.21 Major sites of discovery of Homo erectus. These hominids were the first to migrate out of Africa only about 1 million years ago.

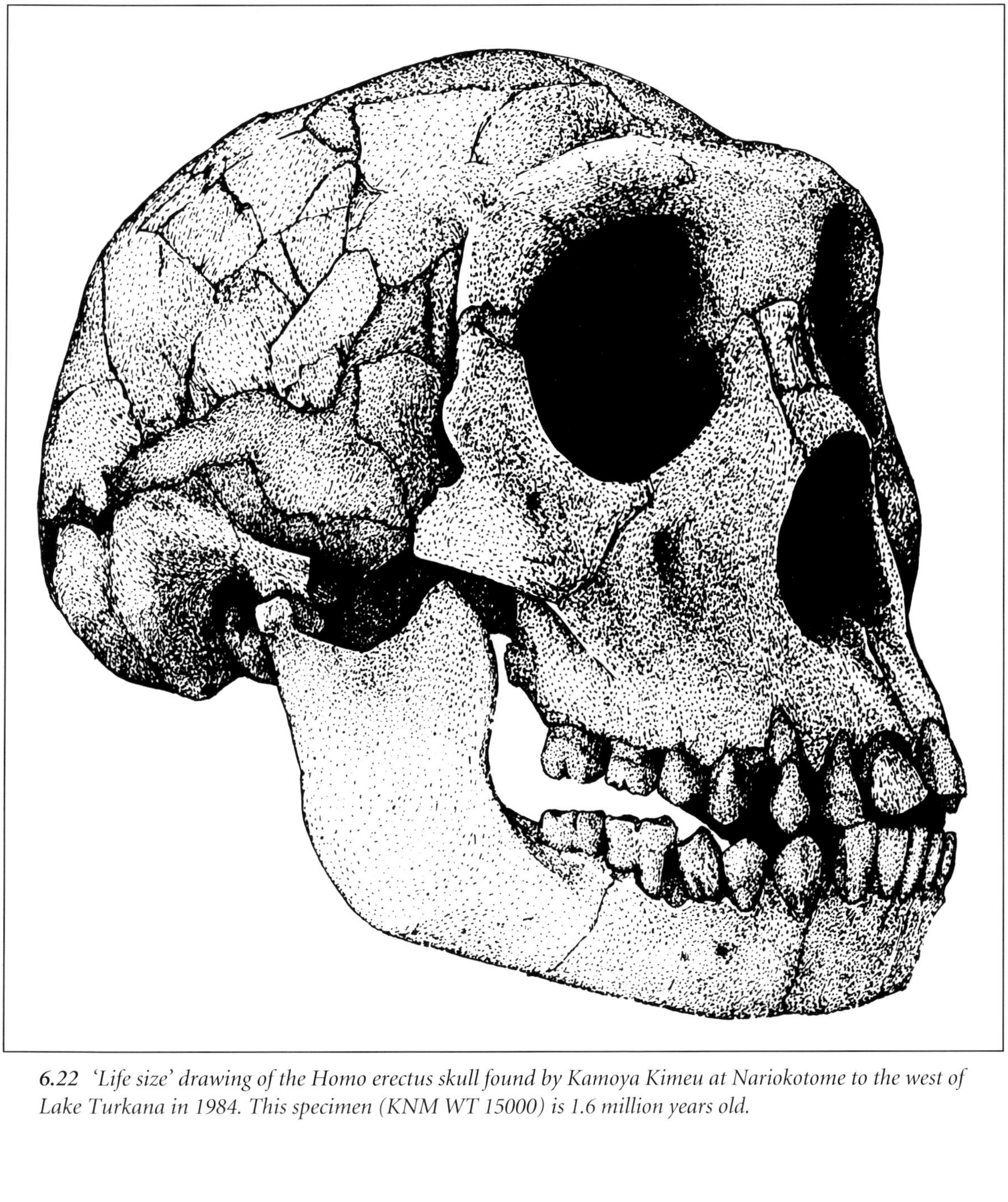

6.22 'Life size' drawing of the Homo erectus skull found by Kamoya Kimeu at Nariokotome to the west of Lake Turkana in 1984. This specimen (KNM WT 15000) is 1.6 million years old.

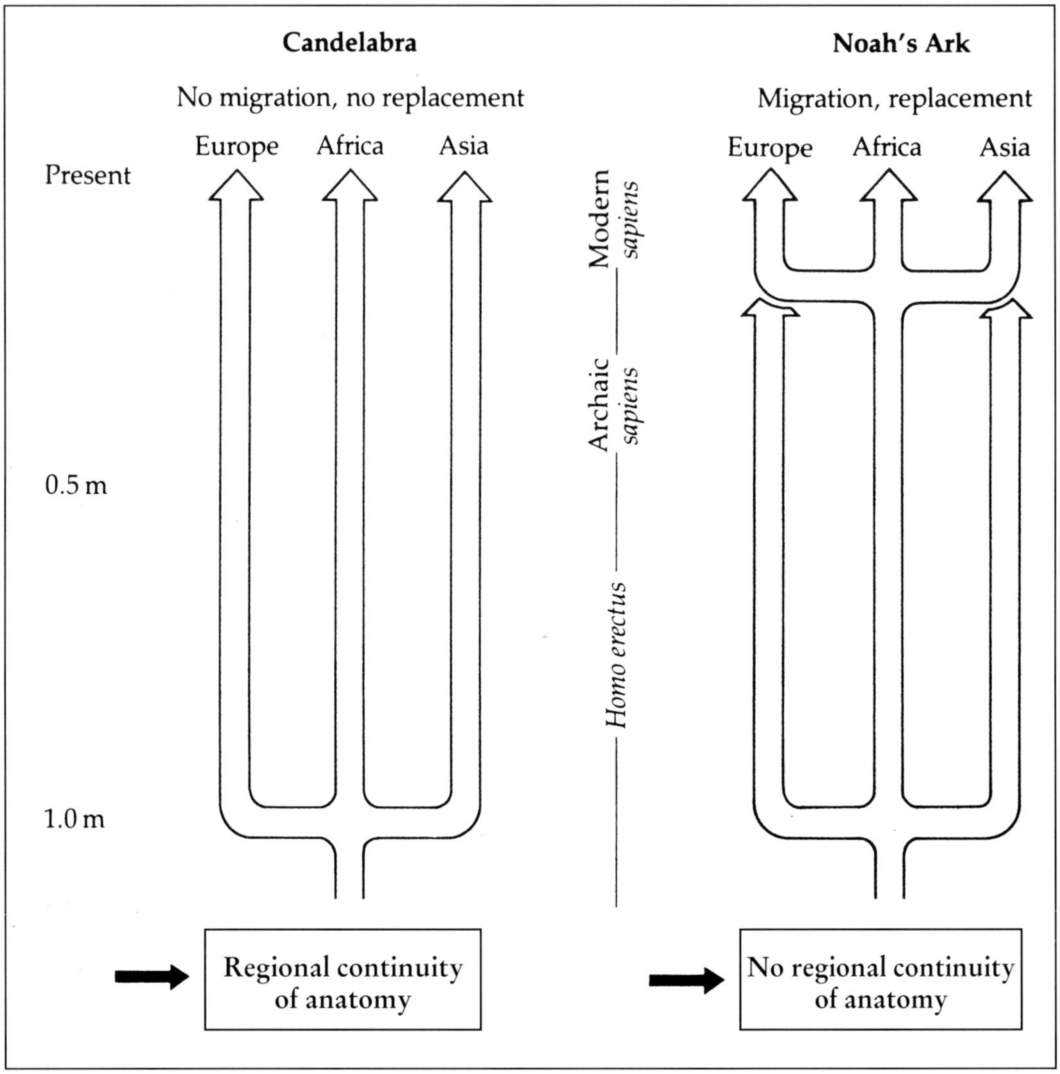

6.23 'Candelabra' and 'Noah's Ark': Two opposite views on the origins of modern humans.

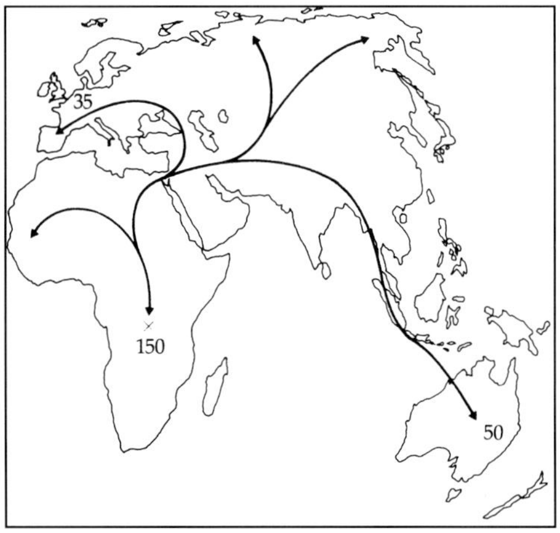

6.24 All the hominid fossils older than about 1 million years were discovered in South- and East Africa. Hominids started to migrate out of Africa only about 1 million years ago, notably Homo erectus.

emerged in Africa between 100,000 and 140,000 years ago. In addition to the indications from molecular biology - notably mitochondrial DNA - fossil evidence supports a model of single origin, followed by extensive migration that started about one million years ago with the *exodus* from Africa. This hypothesis, which William Howells has called the 'Noah's Ark' model, and later has also been named the 'Garden of Eden' hypothesis, opposes the 'Candelabra' model. While the first model assumes subsequent migrations out of Africa of *Homo erectus*, beginning one million years ago, and replaced by recent migrations - equally from Africa - of *Homo sapiens*, the latter, also known as the 'Multiregional' model, regards *Homo sapiens* as the direct, geographical descendant of *Homo erectus* that had moved out of Africa one million years ago.

This ongoing debate expresses the still incomplete fossil record of the period of time in which the evolution processes of *Homo erectus* and *Homo sapiens* have developed. The identification of landmarks such as Dubois' 'Ape-Man from Java', and its congeners 'Heidelberg Man', 'Peking Man', 'Broken Hill Man', and the 'Turkana Boy', have brought the debate further to concentrate now on the origin of modern humans.

While one hundred years ago Dubois started to shift his attention to a new vision of 'Man's Place in Nature' in his ecological studies in the south of Holland, now the work in East Africa has lent a new perspective to the 'Place of our Species', in particular with regard to human's interaction with wildlife communities.

7.1 View of a fen typical of the moorland of South Limburg at the time of Thysse and Dubois (1937).

7 De Bedelaer: New Perspectives on 'Man's Place in Nature'

As we have seen in the previous section on the debate that followed the disclosure of his discovery of the 'Missing Link', Eugène Dubois grew increasingly irritated with scientists and experts who came to study his fossils, and eventually opposed his views.

Since the beginning of his career Dubois had been deeply impressed by the work of Ernst Haeckel. He had almost literally followed his indications as to where in the tropics the remains of early human ancestors might be recovered, *Lemuria*. Also, after his actual discovery on Java, he had adopted Haeckel's original genus name *Pithecanthropus* for his 'Missing Link'. Walther and Heberer in their publication "*Im Banne Ernst Haeckels*"(1953) even described Dubois as:"...filled with enthousiasm for Haeckel's world of thought."

Similarly, as Von Koeningswald noted, Dubois had adopted Haeckel's theoretical views with regard to the processes of phylogenetic evolution. In his letter to Haeckel of 24 December, 1895, Dubois wrote unequivocally: "I should like to tell you how happy I am to be able to express my gratitude for the influence which you, especially through your 'Schöpfungsgeschichte', have exerted on the whole course of my life."

On the other hand, as Bert Theunissen (1989) notes, Dubois indeed took a more practical view than Haeckel, determined to provide more direct, tangible proof from the past to the debate on the human relationship to the apes. He preferred to separate philosophy and science, just as he tried to keep belief and science apart.

Nevertheless, Dubois' growing interest later on in his career in the holistic movement in biology that evolved from embryology and physiology had also been inspired by Haeckel's philosophical ideas. As Peter Bowler (1992) comments on Haeckel's philosophy of monism: "The advocates of such a holistic viewpoint would have been far more willing to accept the need for preserving the complex web of natural relationships, and would thus have gravitated towards the environmentalist movement." By the end of the nineteenth century, the conventional approach to the study of nature had gradually been transformed from a 'natural philosophy' to the new disciplines of biology, genetics, and ecology in all of which different aspects of the environment came to the fore.

It was actually Ernst Haeckel who had introduced the term *Oecologie* to encompass the study of the interactions between organisms and their environment in his *Generelle*

Morphologie of 1866. Derived from the Greek *oikos* that refers to 'family household', he wanted to visualize a kind of 'global household' in which all species interact. In this way, Haeckel's monistic philosophy has certainly promoted a sense of the 'Unity of Nature' among later generations of environmentalists. Darwin, however, was the first to grasp the new science of ecology in all its complexity, especially in view of the rapid changes introduced by man into the environment.

Exactly one century ago again, in 1893, the eminent physiologist J.S. Burdon-Sanderson in his address to the British Association for the Advancement of Science stated that *'oecology'* was one of the three great divisions of biology, the others being physiology and morphology. In his opinion, 'oecology' was " the most attractive of the three because it came closest to the spirit of what had once been called the philosophy of living nature." The modern spelling of the term 'ecology' was established in the same year at the International Botanical Congress.

The new field of ecology, emphasizing the interactions between species and their environment, was also in line with the classical theory of the 'Balance of Nature', proposed by Darwin. Indeed, in his *Origin of Species* (1859), Darwin had stressed the complex interactions with the environment as important factors in determining the selection processes that eventually caused evolution.

7.2 Prehistoric landscape 'Duria antiquior' of Henry de la Besche (1830) visualizing Charles Darwin's conception that: "Old forms are supplanted by new and improved forms".

7.3 Charles Darwin's imaginary landscape of the Second Period (1859).

As Dubois gradually settled into a sedentary life in Limburg, he started to devote more of his time to the study and practice of this new field of ecology. His desire to return to the natural setting of South Limburg had certainly been inspired by the memories of his youth. As we have seen, Eugène Dubois had seized every opportunity in his childhood and later on during his student years to spend as much time as possible in the field.

While he continued to lecture as a Professor of Crystallography, Mineralogy, Geology, and Palaeontology at the University of Amsterdam, he also became Curator of the Geology Department of the Teylers Museum in Haarlem and Director of the *Verzameling Indische Fossielen*, later named *'Collectie Dubois'* at the National Museum of Natural History in Leiden. His collection encompassed about 40,000 fossil floral and faunal remains, brought back from Indonesia in some 400 cases.

It is evident that in addition to his work on *Pithecanthropus erectus* Dubois had adopted a wider scientific perspective on the study of man and nature that included anthropology, biology, anatomy, climatology, and geography. In 1893 he had already published a book on "*The Climates of the Geological Past and their Relation to the Evolution of the Sun*".

In order to pursue his study and experiment with ecology in a more practical setting, in 1906 he acquired the estate *De Bedelaer* near Haelen in Limburg. The estate, on which two fens were situated, also provided Dubois with the *retraite* necessary to be able to study and write, and, as he once noted: "*...accomplish my lifework as well as possible*".

His ultimate objective, however, was to recreate a kind of natural landscape in which early

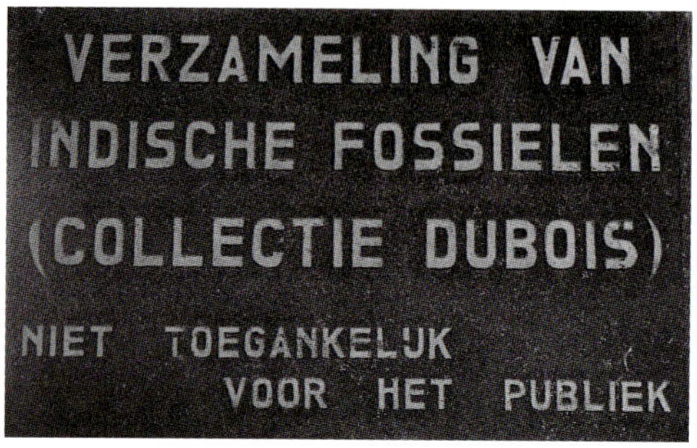

7.4 "Not open to the public." Notice Board at the entrance of the 'Collectie Dubois' at the National Museum of Natural History in Leiden.

7.5 Dr. Leo D. Brongersma, who succeeded Dubois in 1940 as Curator of the 'Collectie Dubois' at the National Museum of Natural History in Leiden.

7.6 Dr. Dick A. Hooijer, who succeeded Dr. Brongersma in 1946 as Curator of the 'Collectie Dubois' at the National Museum of Natural History in Leiden.

7.7 Dr. John de Vos, who succeeded Dr. Hooijer in 1979 as Curator of the 'Collectie Dubois' shows the Pithecanthropus skullcap to Dr. Jan Slikkerveer (May 1993).

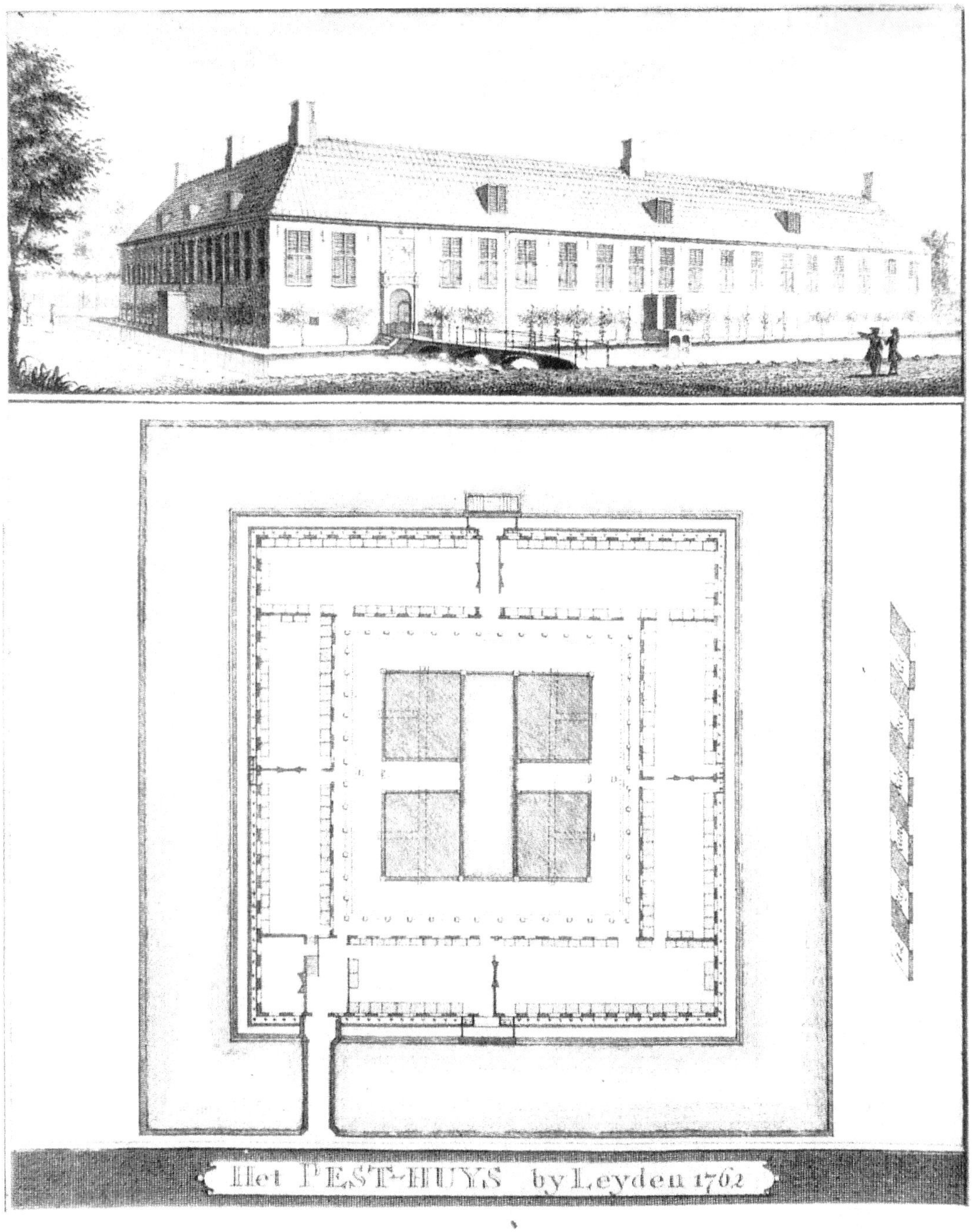

7.8 Plan of 'Het Pesthuys' in Leiden, the hospital for plague patients dating back to 1635, that as a restored monument became the new exhibition premises for the National Museum of Natural History in 1991.

7.9 View of one of the fens of 'De Bedelaer' during the autumn of 1992.

man had lived closely in harmony with his environment. Therefore, upon his return from Java, he devoted the rest of his life to reshaping his estate into a 'prehistoric' nature park, in which many prehistoric but then extinct in Limburg animal and plant species - identified by the fossil record of the area - were to be re-introduced. Using these fossil seeds and other remains he collected from the "Tegelen clay" - deposited during the 'Tiglien Period' - as a guide, he compared these with species still existing today in China and America. Subsequently, Dubois sought to obtain specific seeds and plants for his park from the tropics, in order to reconstruct the vegetation of the times of the earliest Intra-Glacial Era, the Pliocene, in The Netherlands.

As Ton Lemaire (1977-1978) has well documented, initially Dubois was very interested in two issues: the "Tegelen clay" deposits in the area and the origin of the fens. In 1904, he had already made a bore in the clay quarries of the nearby village of Tegelen and was surprised to identify the clay layers as the oldest intraglacial of the Pliocene. The clay contains a wealth of fossil floral and faunal species, then long since disappeared from the temperate zones. The fossil remains indicated a vegetation of exotic forms prevailing in Europe during the Tertiary Period, of which related species still exist today in South-West America and South-East China. The Tegelen fauna in which the warmth-loving plants established the Sino-American or Tertiary floral elements, has been dated to 2 million years ago. Later on, during the 1950s, several pollen analyses of the Tiglien-complex were made confirming the fluctuations in the relative higher temperatures of that period of time (*cf.* Zagwijn 1960).

The faunal fossils also included the remains of a large deer, a horse, a hippopotamus, a rhinoceros, beavers, and even of a species of ape. The fossils clearly indicated a warmer past, during which the "Tegelen clay" had been deposited by the numerous rivers. Dubois spent much time studying these Pleistocene layers and their fossils.

He also studied the collection of fossil faunal and floral remains from the Tegelen area of Dr. Laurens Steijn of Venlo. Later this promted Dubois to have the fossils collected by the labourers from the Tegelen Quarries of the Canoy-Herfkens brick-works in the same city.

The origin of fens, for which several explanations have been produced, was and today still is a controversial issue in geography. In addition to the conventional theory of the involvement of old riverbeds or wind-blown holes, he introduced the hypothesis of water-filled holes or *Sölle,* created by the early accumulation of 'dead' ice from large rivers such as the Maas during the last glacial period.

In the same publication on the geological history of fens, peat moors and dunes, Dubois (1916) already showed himself concerned over the acidification of mist, and rainwater. As he mentions the substantial contribution of smoke from stoves, locomotives, and such things to the production of mist and smog, he explains the high percentage of sulphuric acid in the water of the fens by the many sulphides such smoke contains.

In 1921, Dubois' expertise in hydrology was also called upon by the *Koninklijke Academie van Wetenschappen,* which requested him to draw up a preliminary advice for the Netherlands Government on the state of the spring-water drawn from the dunes near Haarlem in collaboration with the geologist G.A.F. Molengraaff.

7.10 Mansion 'De Bedelaer' in its present state. After its extension in 1972-1973 for the Order of the Réparatricen Sisters it served as a Roman Catholic retreat-house.

7.11 Dubois' drawing of an imaginary landscape near Tegelen at the end of the Pliocene Period. He used fossil finds for his reconstruction of prehistoric animals including the rhinoceros, hippopotamus, and horse.

Jean M. F. Dubois
Naturalist and Explorer
Through the Jungle Land of Guiana
EXPLORING UNKNOWN JAVA AND SUMATRA
EXCLUSIVE MANAGEMENT
GEO. W. BRITT
The Players
120 Boylston Street
Boston

7.12 Dubois' son Jean M.F. Dubois, who travelled around the world as 'naturalist and explorer', sent his father many seeds of exotic plants and trees from the tropics for the 're-vegetation' of 'De Bedelaer'.

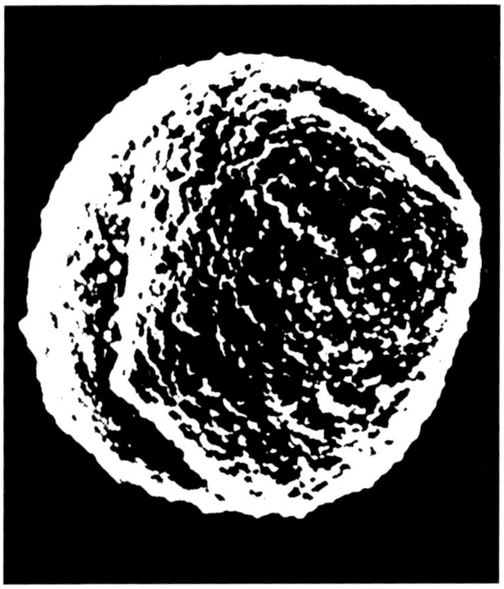

7.13 Electron microscopic image of a pollen of the European oak, retrieved from the 'Tegelen clay' near 'De Bedelaer' (1963).

7.14 The prehistoric elephant that lived about 80,000 years ago in Europe.

His great appreciation of the splendour of Limburg's natural landscape is obvious. In his work he sought to combine scientific knowledge with a feeling for nature: "If we learn a fraction more than just the names of trees, plants, and birds, we will meet old friends, not strangers." Equally, in his practical experiments in geology and hydrology on his estate, Dubois expressed his hope to create more general awareness of the beauty of nature.

When Dubois bought the estate on 20 June, 1906, it was described as De *Groote Bedelaer* including 13.5 hectares of *"dennenbosch, water en heide* " (fir-wood, water, and moorland), and about 6 hectares of *"berkenbosch en hakhout* "(birch-wood and coppice). In the course of the following years he was able to extend the estate to 38 hectares in 1937 by acquiring additional areas of land. In contrast to the present situation with its rich and varied vegetation, the original state, in Dubois' opinion, was fairly poor, not to say 'desolate'. The area included the two fens, *De Groote Bedelaer* and *De Kleine Bedelaer.*

As he planned to change this situation as soon as possible, as a scientist he first began by making a complete inventory of its vegetation. This inventory, carefully recorded in one of his six diaries and named '*Kallilimne*' (Greek for 'beautiful lake'), was made with the assistance of the renowned botanist Dr. L. Vuyck, who was then the Director of the *Nederlandsche Koloniale Landbouwschool* (Netherlands Colonial Agricultural College) in Deventer. The original list of 80 plant names, *"Flora of Kallilimne"*, is dated 23 July, 1907.

7.15 Detailed map of the site of 'De Bedelaer' near Haelen in Limburg (1850).

This list was later extended by Dubois himself by the addition of six names of herbs which he had planted as a first phase in the large vegetation plan.

Based on Dubois' inventory, an extensive reconstruction of the original structure of the vegetation in 1907 was made in 1977 by the plant sociologist Prof. V. Westhoff in collaboration with the anthropologist Ton Lemaire. Their reconstruction, encompassing the fen itself, its banks and its immediate environment, shows a relatively poor moorland fen, as was common at the beginning of this century in Limburg. Nowadays, such conditions are only found in a few natural reserves in the south of Holland. Several plant species were recorded, which then in 1977 about seventy years later, had become particularly rare. These include *Scheuchzeria* and *Malaxis.*

In 1908 and 1912, Dubois also studied the water of the fen itself. Several specimen were taken by two experts. Their reports are still in the Archives of The Netherlands Hydrobiological Association. According to the analysis by the Curator, Drs. P. Schroevers, the condition of the water at the time was interesting: rich in species as a result of processes of transition. In 1977, a fen in such condition would certainly have been declared as a 'protected area'. Dubois used the reports for the development of his theory on the origin of fens in the low countries.

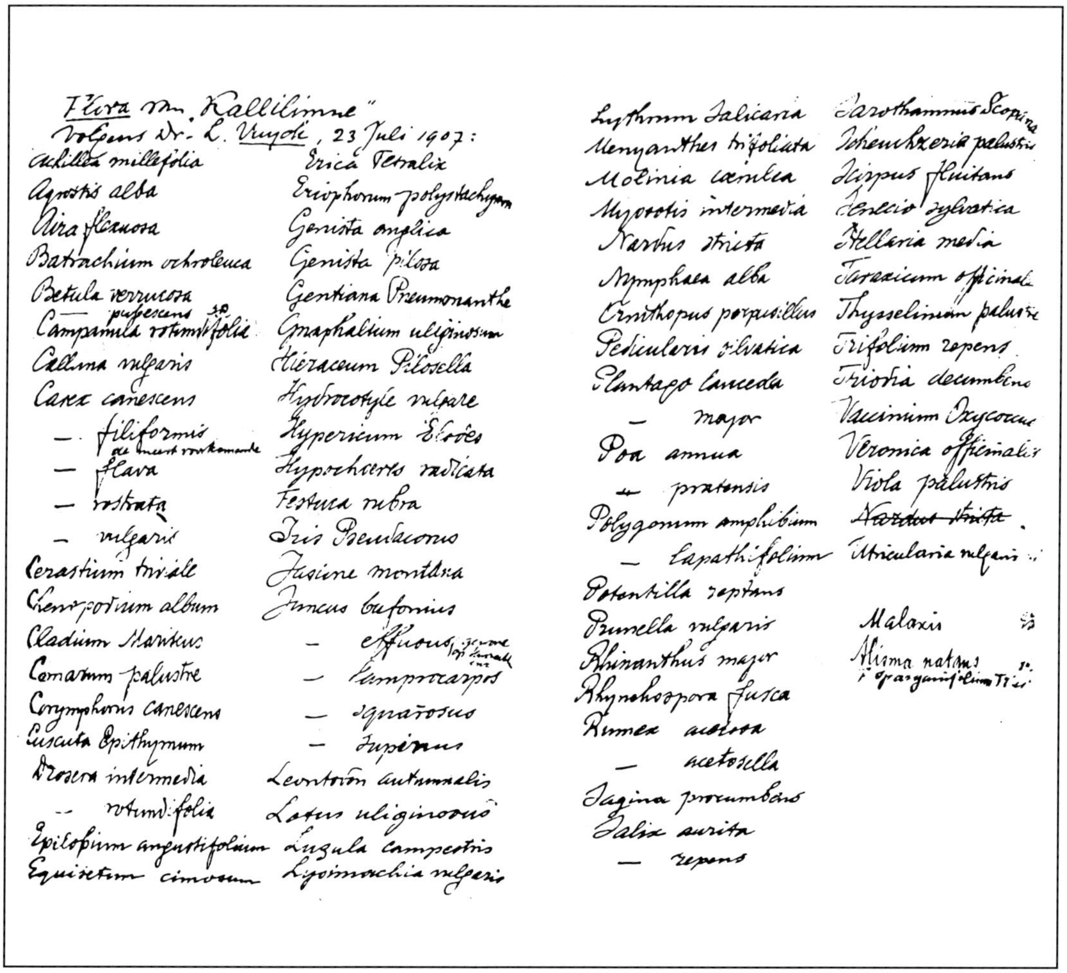

Flora m. Kallilimne"
volgens Dr. L. Vuyck, 23 Juli 1907:

Achillea millefolia
Agrostis alba
Aira flexuosa
Batrachium ochroleuca
Betula verrucosa
— pubescens
Campanula rotundifolia
Calluna vulgaris
Carex canescens
— filiformis
— flava
— rostrata
— vulgaris
Cerastium triviale
Chenopodium album
Cladium Mariscus
Comarum palustre
Corynephorus canescens
Cuscuta Epithymum
Drosera intermedia
— rotundifolia
Epilobium angustifolium
Equisetum limosum

Erica Tetralix
Eriophorum polystachyum
Genista anglica
Genista pilosa
Gentiana Pneumonanthe
Gnaphalium uliginosum
Hieracium Pilosella
Hydrocotyle vulgare
Hypericum Elodes
Hypochaeris radicata
Festuca rubra
Iris Pseudacorus
Jasione montana
Juncus bufonius
— effusus
— lamprocarpos
— squarrosus
— supinus
Leontodon autumnalis
Lotus uliginosus
Luzula campestris
Lysimachia vulgaris

Lythrum Salicaria
Menyanthes trifoliata
Molinia caerulea
Myosotis intermedia
Nardus stricta
Nymphaea alba
Ornithopus perpusillus
Pedicularis silvatica
Plantago lanceolata
— major
Poa annua
— pratensis
Polygonum amphibium
— lapathifolium
Potentilla reptans
Prunella vulgaris
Rhinanthus major
Rhynchospora fusca
Rumex acetosa
— acetosella
Sagina procumbens
Salix aurita
— repens

Sarothamnus Scoparius
Scheuchzeria palustris
Scirpus fluitans
Senecio sylvatica
Stellaria media
Taraxacum officinale
Thysselinum palustre
Trifolium repens
Triodia decumbens
Vaccinium Oxycoccus
Veronica officinalis
Viola palustris
~~Nardus stricta~~
Utricularia vulgaris

Malaxis
Alisma natans

7.16 List of Plant Names of the fen at 'De Bedelaer' from Dubois' diary of 1907.

7.17 Cranberry Marsh Whortleberry (Oxycoccus palustris 383), mentioned in Dubois' List of Plant Names of 1907 as Vaccinium palustris.

7.18 Long-Leaved Sundew (Drosera longifolia) mentioned in Dubois' List of Plant Names of 1907.

7.19 Least Bog Orchid (Malaxis paludosa, 654) mentioned in Dubois' List of Plant Names of 1907.

7.20 Marsh Scheuchzeria or Rannoch Rush (Scheuchzeria palustris L 1368) mentioned in Dubois' List of Plant Names of 1907.

7.21 White Water-lily (Nymphaca alba 539) mentioned in Dubois' List of Plant Names of 1907.

7.22 Reconstruction of the Carboniferous Stigmaria tree, known from the Limburg fossils, drawn by Eli Heimans (1911).

Since his objective had been to reshape his estate into a *Natuurpark* (Nature Park), Dubois recorded all his experiments, measurements, and analyses in his *Kallilimne Nota,* supplemented with excerpts from several works on the creation of gardens and parks such as the famous *Gartenbuch* of Hampel. The report of a visit by the *Limburgsch Natuurhistorisch Genootschap* (Limburg Society for Natural History) in 1938 reads that Dubois': "*...primary objective was the establishment of a natural reserve or a nature park, in which plants and animals could live at liberty*".

To this end, he cultivated a so-called '*Vogelboschje*' (bird thicket), designed to establish a bird sanctuary on *De Bedelaer.*

Dubois performed three types of interventions in order to realize his ambitious plans: a) management of the water level; b) fertilization of water and soil; and c) large-scale plantation of trees and shrubs. Dubois' idea was that earlier deforestation had caused the water level to rise. As he wanted to lower the actual level by one metre in order to arrive at the 'original' level, he developed a method of draining the water through a vertical pipe, 150 mm wide by 17 m. long through the deep layer of gravel into the bedding of the River Maas.

After the level of the water - to the great surprise of the local farmers - was indeed lowered by 1 m., he started to fertilize the water and the surrounding soils. Large quantities of minerals such as ground iron (100 kg.), chalk, and marl (2500 kg. and twice 3000 kg.), phosphate (1300 kg.), and so forth were added to the water between 1907 and 1913, which resulted in a dramatic change in vegetation. Various species of fish were also introduced in 1914: 1000 rudds; 600 Masurian tenches; and smaller quantities of other species.

7.23 Reconstruction of the prehistoric landscape of Limburg, based on the fossils from the Tegelen quarries in Limburg. The autumn forest includes the typical Pterocarya tree.

7.24 The Marsh Thistle (Cirsium palustre) is still common in the wet marshy areas of Holland (1992).

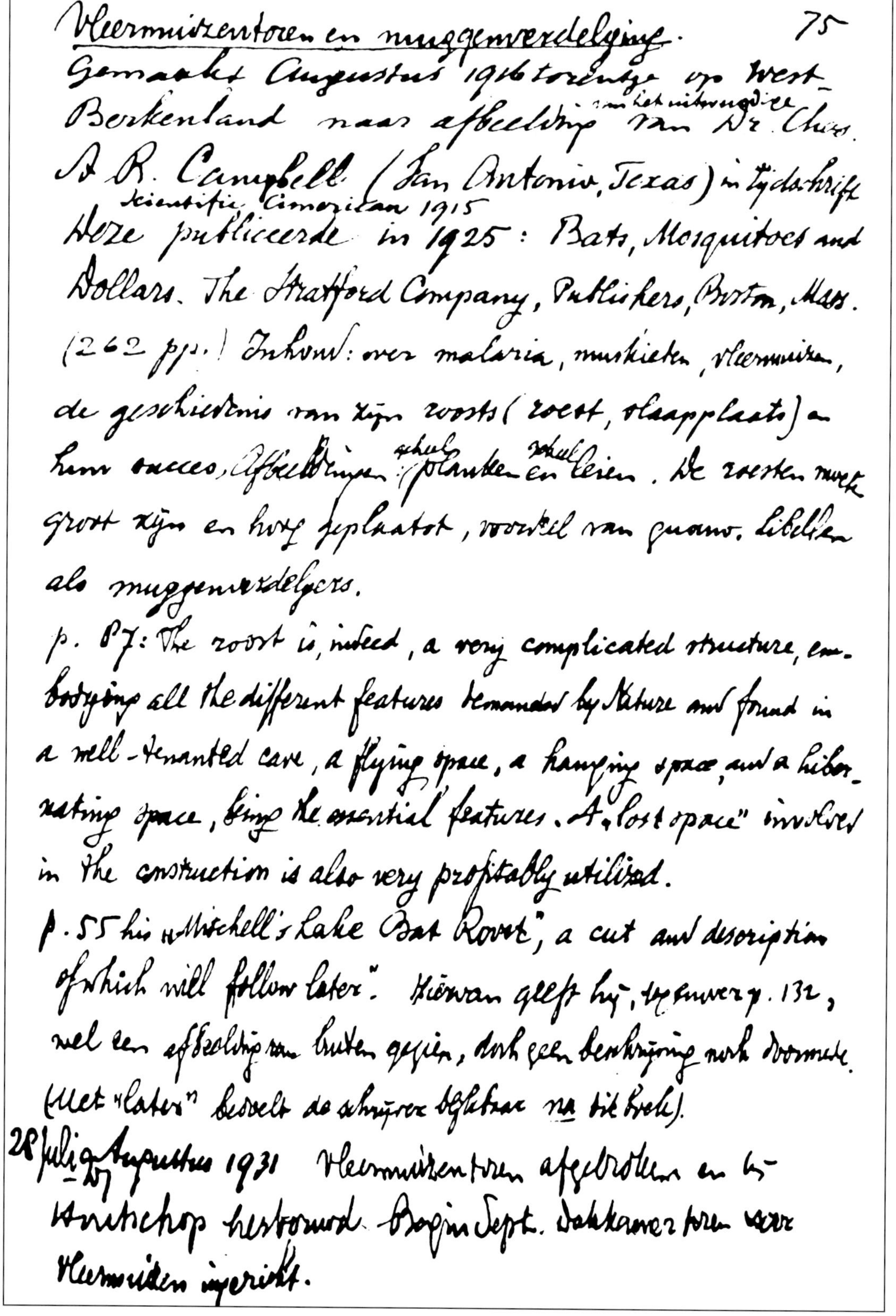
Vleermuizentoren en muggenverdelging. 75
Gemaakt Augustus 1916 torentje op West-Berkenland naar afbeelding met het uitwendige van Dr. Chas. A. R. Campbell (San Antonio, Texas) in tijdschrift Scientific American 1915
Deze publiceerde in 1925: Bats, Mosquitoes and Dollars. The Stratford Company, Publishers, Boston, Mass. (262 pp.) Inhoud: over malaria, muskieten, vleermuizen, de geschiedenis van zijn roosts (roost, slaapplaats) en hun succes, Afbeeldingen schuil-planken en schuil-lieren. De roosten moeten groot zijn en hoog geplaatst, voordeel van guano. Libellen als muggenverdelgers.
p. 87: The roost is, indeed, a very complicated structure, embodying all the different features demanded by Nature and found in a well-tenanted cave, a flying space, a hanging space, and a hibernating space, being the essential features. A „lost space" involved in the construction is also very profitably utilized.
p. 55 his „Mitchell's Lake Bat Roost", a cut and description of which will follow later". Hiervan geeft hij, tegenover p. 132, wel een afbeelding van buiten gezien, doch geen beschrijving noch doorsnede. (Met „later" bedoelt de schrijver blijkbaar na dit boek).
28 juli–27 Augustus 1931 Vleermuizentoren afgebroken en tot houtschop hervormd. Begin Sept. dakkamertoren voor vleermuizen ingericht.

7.25 Dubois' 1931 notes on his experience with the 'bat-roost' he built on 'De Bedelaer' in 1916 after C.A.R. Campbell's model in Texas.

7.26 'Bat-tower' built by Dubois on 'De Bedelaer' in 1916, and restored by the 'Belgisch-Nederlandse Vereniging voor Zoogdierkunde en Zoogdierbescherming' during the 1980s.

7.27 Photograph taken of Eugène Dubois on 'De Bedelaer', shortly before his death in 1940.

7.28 Dubois had received many rewards in recognition of his scientific work, including a Royal Decoration by the end of his life.

The soils of the estate were also fertilized after tilling. In addition to the plantation of many smaller plant species, he started to introduce large numbers of different types of shrubs and trees.

Among the trees which Dubois had re-introduced into his park were various exotic shrubs known from the Tegelen fossil record, such as the Trans-Caucasian ornamental tree *Pterocaya*, the tulip-tree (*Magnolia cor*), and the Chinese rubber tree (*Eucommia actinidia*). Most were brought in from other countries, sometimes even from other continents. In this way he also imported the unique mammoth tree (*Sequoia gigantea*) and the swamp cypress (*Taxodium distichum*) from North America, still flourishing there today. The great variety of coniferae and deciduous trees were represented by species and sub-species, such as *e.g.* the oaktree (*Quercus*) with its 9 subspecies, 7 of which also originate from North America. He also brought in exotic species that were not found in the fossil record of the area.

Throughout his work on *De Bedelaer*, Dubois opposed the rational exploitation of forests, then being promoted by German forestry practices. By the end of the nineteenth century, a more nature-oriented mode of forestry was introduced. Dubois' holistic approach included the creation of a highly varied, 'mixed' forest, combined with respect for the aesthetics and splendour of the forest.

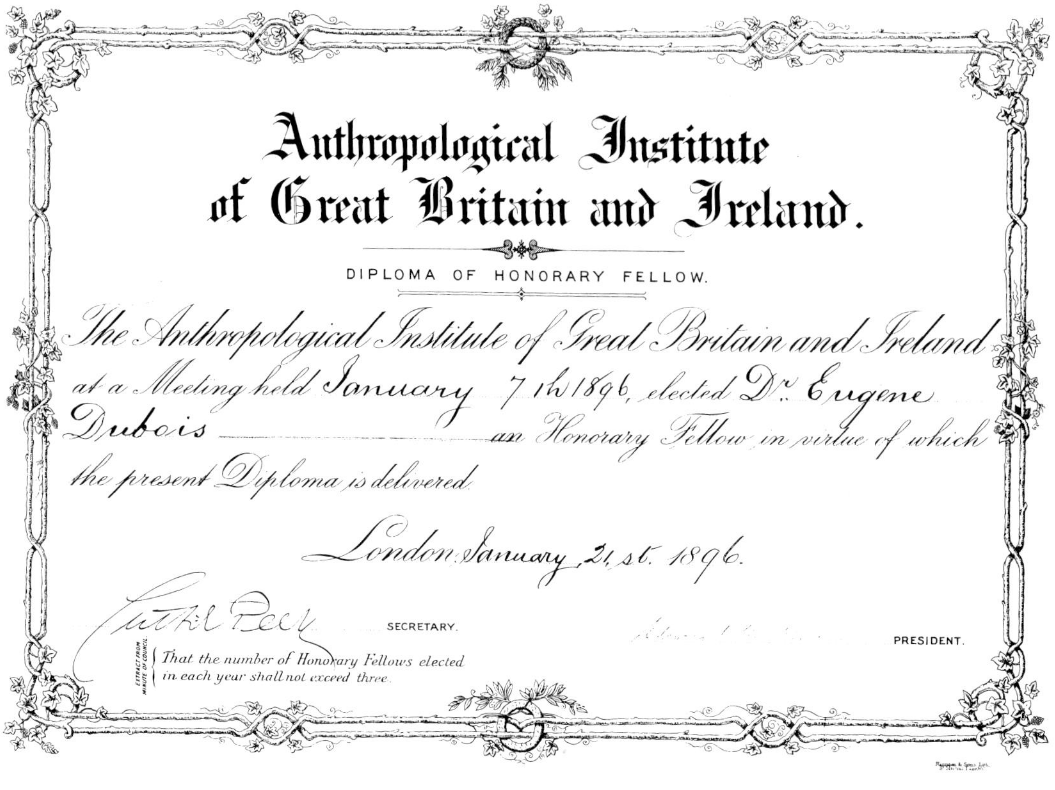
Anthropological Institute
of Great Britain and Ireland.

DIPLOMA OF HONORARY FELLOW.

The Anthropological Institute of Great Britain and Ireland at a Meeting held January 7th 1896, elected Dr. Eugene Dubois an Honorary Fellow in virtue of which the present Diploma is delivered.

London January 21st 1896.

SECRETARY. PRESIDENT.

EXTRACT FROM MINUTE OF COUNCIL: That the number of Honorary Fellows elected in each year shall not exceed three.

7.29 In addition to the Honorary Fellowship of the Anthropological Institute of Great Britain and Ireland of 1896, Dubois received an Honorary Doctorate from the University of Amsterdam in 1897.

Complementing his fossil evidence approach and his 'law of cephalization' as entirely new means to substantiate the theory of human's direct relation to nature, his practical fieldwork in the south of Holland finally took shape. As a true naturalist he recreated *De Bedelaer* as a prehistoric landscape in Limburg that served to make our past even more tangible and relevant to our understanding of the further evolution of *Homo sapiens.*

In June 1978, Prof. Westhoff made an inventory of the vegetation of the estate at that point in time, and compared it with the inventory made by Dubois about seventy years ago. His conclusion was that most of the original vegetation, characteristic of fens, had disappeared to make way for a variety of exotic trees and shrubs. Although nowadays one might object to the planned intervention into an environment that with the passing of time has become fairly rare, the concept of 'untouched nature' - as Ton Lemaire observes - had just begun to develop within a small circle of naturalists and conservationists in Dubois' own time.

Of special interest is the fact that Dubois erected a few 'bat-roosts' or 'bat-towers' on his estate after Charles H.R. Campbell's model in San Antonio, Texas. In 1916, inspired by the publication of Campbell's results regarding the eradication of malaria mosquitos near Mitschell's Lake in *Scientific American* (1915), he built a small wooden tower and a larger brick tower, in which a stove for the winter was established. His primary objective was to

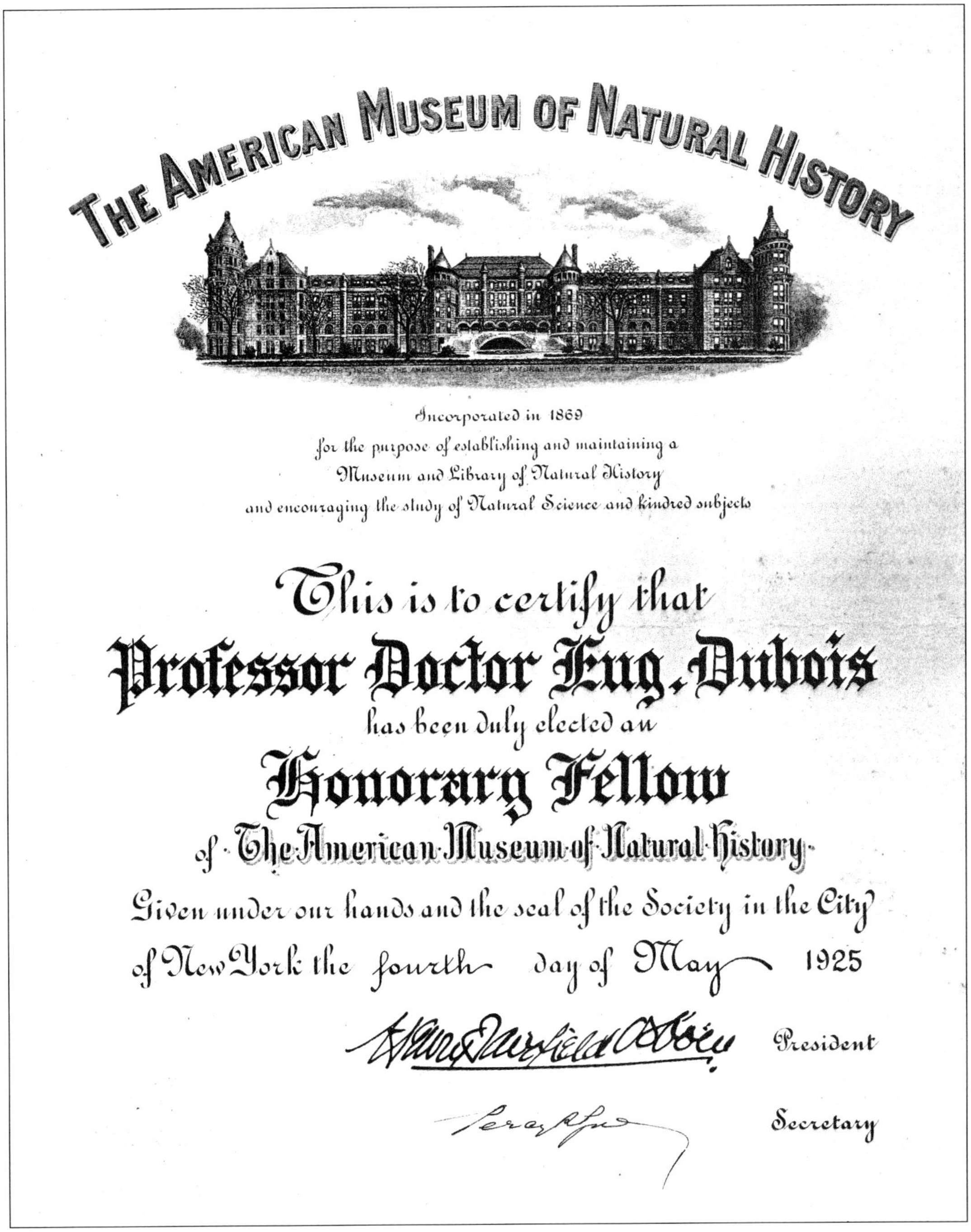

7.30 *In 1925, Dubois received also an Honorary Fellowship of the American Museum of Natural History in New York, U.S.A.*

establish a colony of bats in line with his efforts to secure the floral and faunal diversity in his '*natuurpark*'. However, by so doing he also introduced a kind of biological method for the eradication of mosquitos living in the fens of *De Bedelaer*. Although nowadays the construction of bat boxes is regarded a more appropriate way for bat protection, Dubois' two remaining 'bat-towers' constitute a remarkable monument to bat conservation from the past. Following an article by A.M. Voûte and P.H.C. Lina on Dubois' 'bat-towers' in the *Natuurhistorisch Maandblad* (1983), the *Belgisch Nederlandse Vereniging voor Zoogdierkunde en Zoogdierbescherming* (Belgian-Netherlands Society for Zoology and Mammal Protection) launched a successful rescue operation for the restoration of the wooden 'bat-tower' on *De Bedelaer*.

The *Nederlandsche Vereeniging tot Behoud van Natuurmonumenten* (Netherlands Society for Nature Conservancy) had just been established through the initiative of Dr. Jac P. Thysse, the Founding Father of the Netherlands movement for the conservation of nature. The new concept of '*Natuurmonument*' ('Monument of Nature') was then introduced to keep and conserve certain parts of The Netherlands as a 'monument' to the past: "... *in order to portray a part of Holland's origin that never could be realized in a museum or by a description*" (Vuyck 1901).

While his friend Thysse preferred to maintain and conserve the indigenous habitat, Dubois believed, that through intervention and management, a valuable landscape from the past could be recreated. By means of a combination of forestry and conservation, he established a nature park that did indeed come close to portraying the early prehistorical landscape of Limburg. Although to begin with Thysse's approach tended to dominate nature conservation efforts in Holland, new insights into vegetation science have recently resulted in certain modifications that are more in line with Dubois' views.

As a natural scientist Dubois certainly played an important role in the Netherlands movement for the conservation of nature. From 1912 he was a Member of the Board of both the Netherlands as well as of the International Society of Nature Conservancy. He reaped a lot of fame for the conservation of nature, and in a letter on the occasion of Dubois' 80th birthday, Thysse praised his great achievements in this field.

7.31 Dubois died in 1940 on his estate 'De Bedelaer', and was buried in unconsecrated ground in the nearby city of Venlo. On his tombstone are the skullcap and (two!) crossed femurs of Pithecanthropus.

The General Assembly of the *Nederlandsche Vereeniging tot Behoud van Natuurmonumenten* when commemorating his birthday, also noted: "This world-famous natural scientist has been a Member of our Board for many years and devotes his great influence and his eloquence specifically to supporting the International Conservation of Nature and to introducing this movement into the scientific world in order to open the eyes of those who do not sufficiently recognize what an important loss would be cognize what an important loss would be suffered by mankind if fauna and flora, particularly of the unspoiled areas were to disappear."

The Assembly continued to underscore Dubois' own vision that such loss would indeed be: "...a loss of forms of life, and precisely those that should teach us most about the essence of life itself, and such a loss would indeed be the most dramatic that can afflict the human spirit." Furthermore, the Assembly notes that in Dubois': "*...terse assertion lies the foundation of the Conservation of Nature and as such it more or less expresses the objective of our Society.*"

When Dubois died on his beloved estate *De Bedelaer* on 16 December, 1940, at the age of almost 83 years, many articles and messages throughout the country commemorated the passing of this internationally renowned scientist. Understandably, the emphasis was laid on his discovery of the *Pithecanthropus erectus*, the Ape-Man from Java.

However, as has also transpired from Ton Lemaire's studies (1977-1978), so far the true significance of Dubois work and vision has remained rather outside the limelight: his evidence of a new perspective on our place *within* - and not *above* - nature.

7.32 a + b The Dutch naturalist and author Eli Heimans became a regular visitor to 'De Bedelaer', where he joined Dubois in his ecological studies (August 1936).

At his death, Dr. Van Tienhoven delivered a commemorative address on Eugène Dubois during the General Assembly of the Netherlands Society of Nature Conservancy on 15 March, 1941, which then already portrayed Dubois' life and works in a wider context. He stated, that:" *His* (Dubois') *pronouncement that the extinction of the large animal species and of the significant flora would be one of the greatest losses for the whole of mankind has opened the eyes of so many.*"

His estate *De Bedelaer* had gradually transformed into Holland's first 'nature park', and attracted visits and excursions from leading naturalists such as Jac. P. Thysse and Eli Heimans and associations including the Society of Natural History of Limburg and the Society for Nature Conservancy.

Even today, *De Bedelaer* dominates the Limburg landscape as a unique monument to Dubois' scholarly work and vision and this at a time when the newly-developing field of study of our place in nature from prehistoric times up to the present is gaining fresh impetus.

8.1 *The chimpanzee 'Freud'. Gombe National Park, Tanzania (1992).*

8. The Past is the Key to Our Future

A century ago, Eugène Dubois initiated 'fossil hunting' in Asia, notably in the former Netherlands East Indies in order to substantiate the relationship between man and his closest relative in the animal world, the ape. Later on, the search for human origins was virtually revolutionized by Mary and Louis Leakey with their world-famous discoveries in Africa. The focus of research has now shifted not only to East Africa, which Charles Darwin had called the 'Cradle of Mankind', but more recently a wider context, an extra dimension, has been added to the research into human ancestry. It involves man's need to understand the nature of humanity and our place in the world.

With this development, palaeoanthropology, basically the study of the history of man's place in nature has gradually begun to encompass philosophical and metaphysical aspects of the story of mankind, *i.e.* of ourselves. Over the past hundred years or so, the reconstruction of the human 'family-tree' has revealed a process of origin, transformation, and survival of one specific human-like species, *Homo sapiens sapiens*, modern man. However, the study of this latest offspring of the human tree goes beyond the scientific analysis of fossil remains, molecular data, and cultural characteristics; it needs the understanding of 'extrascientific' aspects of what has made us 'human'.

Indeed, current questions such as how, where, and when did modern humans evolve and what is 'humanness', are demanding a new perspective on the place of Modern Man in the universe of things.

Such a significant new perspective on the place of our species, which had already occupied Dubois' mind after his discovery of an early ancestor in Java a century ago, has taken further shape in Africa. The fascinating but tenuous efforts by which the parts of this evolutionary puzzle are slowly being pieced together has recently been further documented in *Origins Reconsidered. In Search of What Makes Us Human* (Leakey and Lewin 1992).

On the one hand, such a perspective is firmly grounded in recent developments in the sciences directly involved in the study of evolution. On the other hand, it also derives from new approaches to what has lately been called the 'environmental sciences'.

The first developments included - as we have seen - the new, spectacular discoveries at

8.2 Chimpanzees are, as all primates, highly social animals. 'Fifi' and her 10-day old son 'Ferdinand' in Gombe National Park, Tanzania (1992).

Lake Turkana of *Australopithecus, Homo habilis,* and *Homo erectus,* and the related first migration out of Africa, starting about 1 million years ago. This was soon followed by the late Glynn Isaac's evidence on early food-sharing communities which rendered previous views of Man the Hunter almost untenable (1984). This side of the new perspective has recently been extended by a reinterpretation of the origins of modern humans by looking at the 'fate' of the Neanderthals. Although the 'Noah's Ark' hypothesis on migration replacement has so far lacked sufficient fossil evidence, it seems to be supported by the analysis of patterns of genetic variation of mitochondrial DNA among modern human populations.

Additional developments in the recent study of early language and communication, and the related study of fossil hominid brains and vocal apparatus, including the larynx, indicate that *Homo erectus* already had a rudimentary spoken language. In this context, it is interesting to note that as early as in 1886 Dubois had been involved in larynx research. In 1886, prior to his journey to The Netherlands East Indies, he published two articles on respectively the human larynx and on the larynx in whales.

Furthermore, results from comparative primate studies have provided significant information on the social structure of human ancestors and on the primate mind. These include the results of two world-famous primate field studies established by Louis Leakey: Jane Goodall's

8.3 a Akela trying to imitate drawing in Marvin Egberts' notebook at the Jane Goodall Institute, Bujumbura, Burundi (1992).

8.3 b Akela's drawing 'bridging the gap' between Man and Ape (1992).

Chimpanzee Centre in Gombe, Tanzania, and Diane Fossey's Gorilla Sanctuary in Rwanda. More insight has been gained into the role of complex social interactions and alliance networks among individuals of specific primate groups.

Recent focus in primatology on related ecological and behavioural aspects has brought the definition of 'primate' alongside that of 'human' closer to our understanding of evolutionary processes of behavioural ecology.

The analysis of the evolution of human consciousness leads to the assumption that, as a 'quality of mind', it has fostered Modern Man himself "to feel rather special in the world, separate from and somehow above the rest of nature".

Finally, three fascinating sub-areas of concern are elaborated on in *Origins Reconsidered* (1992) in relation to this new perspective of the place of our species: Inevitability; the Gap; and the Sixth Extinction.

While the first concept refers to the coincidental character of human evolutionary history rather than a predestined 'path', the second idea of the 'gap' between humans and the rest of the animate world is unmistakably shown to be: "something of an illusion, an accident of history." The evidence of the fossil record, initiated in Asia by Dubois and since then vastly extended elsewhere, notably in Africa, relates us directly to the rest of nature.

***8.4** 'Chief' Rugabo, the Silverback Mountain Gorilla in the Virunga Mountains of Zaire (1992).*

8.5 'Maheshe' the Mountain Gorilla of the Kahuzi Biega Park depicted on a banknote of Zaire (1991).

8.6 Food-sharing for survival. Two male lions with their antelope prey in Virunga Park, Zaire (1992).

8.7 Freud in his favourite 'nishe' of the top of an oilpalm tree in Gombe National Park, Tanzania (1992).

The third concern - the Sixth Extinction - is related to the other side of the new perspective on the place of our species: it deals with present *Homo sapiens* and his prospects for the future.

In the interaction between human populations and the natural world, a process of rapid decline of the animal and plant kingdom has reached a point at which the world's species will probably be halved within a period of three decades. Due to increased population growth and economic development, existing biodiversity is eroding fastly. Although such a threatening process of extinction is not entirely new to planet Earth, this time its dimensions seem different.

The history of our planet has seen five mass extinctions since the earliest times of the origin of complex forms of life, during which living species were almost completely wiped out: the Ordovician (430 millions years ago); the Devonian (350 million years ago); the Permian (225 million years ago); the Triassic (200 million years ago); and the Cretaceous (65 million years ago). Despite such mass incidents of extinction, periods of recovery, in which surviving species used ecological opportunities to restore levels of diversity occurred rapidly.

In such a sequence of a long-term oscillating process of extinction and recovery in the history of life on Earth, we are presently about halfway into the Sixth Extinction, a loss of 50 per cent of species. This time there will be no natural catastrophe, but the impact of an exponential population growth threatening the entire ecosystem. The fossil record has revealed that, generally, species have a longevity of about 5 - 10 million years; vertebrates only 2 million

8.8 *Indonesia's efforts to protect the Orang-utan of Sumatra. Special issue of stamps on the occasion of the International Conference on the Largest Ape in the World.*

8.9 *Special stamp from Uganda depicting the extinct 'Hypsilophodon', a bird-hipped creature from the Mesozoic Era.*

8.10 G.H. Schubert's (1882) illustration of two major endangered species: the rhinoceros and the elephant.

8.11 *Since its formation, Kenya Wildlife Service has caught the world's attention by burning confiscated ivory worth 30 million $ to ban the trade (1989).*

8.12 *The monument that was erected in Nairobi National Park to commemorate President Daniel Arap Moi's well-publicized burning of confiscated ivory in 1989, with Dr. Richard Leakey, Director of Kenya Wildlife Service.*

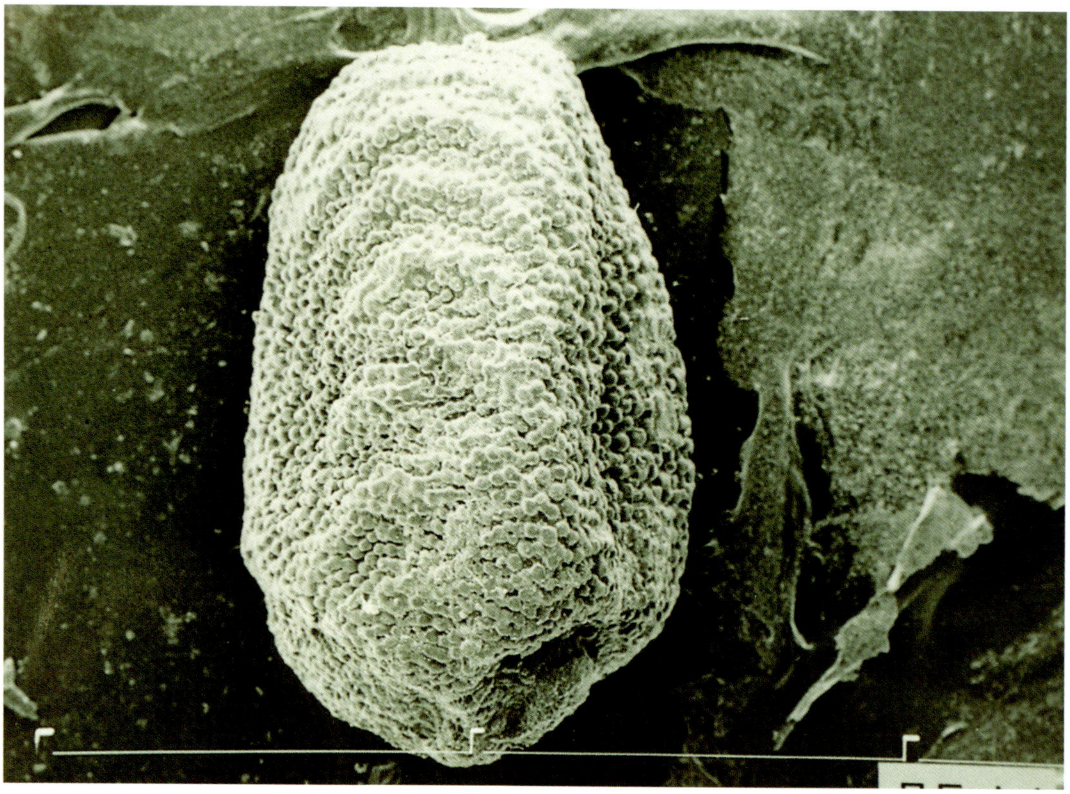

8.13 Electron microscope photograph of Elusine indica ssp. africana, a progenitor of finger millet, retrieved from Gogo Falls, West Kenya. It confirms Vavilov's assumption of Eastern Africa's early 'Centre of Biodiversity'.

years. Although our species in this respect is very young - about 100,000 years old - the prospects do not look good. Even if it is not by self-destruction in the short term, "*...we still face the prospect that there will one day be an Earth with no Homo sapiens on it* (*cf.* Leakey and Lewin 1992).

After the completion of the human-inflicted Sixth Extinction, perhaps some 700 million years after the origin of complex forms of life on Earth, certain species will again survive to continue the process of alternating phases of extinction and recovery. However, the study of the fossil record has shown that the human species is part of - *not apart from* - the natural world on Earth, here today. The fossil record has also provided us with the knowledge that a species, once extinct, is destroyed forever, and that destruction of the environment will eventually pertain to the reduction of life.

The specific qualities acquired in the course of the evolutionary process by *Homo sapiens* - previously referred to as 'humanness' - has provided us not only with an insight into our origins, but also with a responsibility towards other species. The development of 'culture' that has profoundly transformed the life of *Homo sapiens* could perhaps help to anticipate our immediate future in the short-term. But, as biodiversity is threatened with extinction, so too it seems is cultural diversity in terms of human's various traditional systems of knowledge and technology.

8.14 The French odontologist Pierre-Fr. Puech's image of 'Human Evolution in its Ecological Context'.

As our 'stewardship' of Earth has to change into 'tenancy', and as the study of humans' relationship with nature gradually has to transform into that of human's interaction with the natural world, the insight into the prevailing historical processes seems indispensable. Probably in the same way as anthropologists and archaeologists have collaborated to understand Man's development process since the earliest period of prehistory to the present day by using the method of 'ethnographic analogy', should we reconsider our age-long interaction with the living world over a longer period of time.

Contemporary complex configurations of various population groups within their ecosystems could certainly be better unravelled by (pre)historical analysis. The specific knowledge of those groups depending more directly on their immediate environment in this respect has proved to be most critical for survival.

An example of the daily practice of wildlife conservation and protection in the area where humans most probably originated, East Africa, particularly in Kenya is that in the distant past indigenous population groups and wildlife have co-existed for a long time in a rather balanced way. In view of the increasing complex situation due to factors such as population pressure and shortage of land such local, indigenous experience and knowledge could be re-assessed and revitalized, and eventually incorporated into an overall wildlife management policy planning and implemention programme. With regard to the contemporary rapid extinction of particular species, the experience is that not all such detrimental processes result from mere animal-human competition, but rather from fatal impacts such as excessive game hunting and poaching.

Serious threats to the survival of endangered species such as African rhinos and elephants, for example, could only be taken away by joint international efforts such as the Convention of International Trade in Endangered Species (CITES) and specific efforts such as Kenya's firm stand on the ivory ban. Revitalization of such a model of population - wildlife communities' interaction such as through the introduction of a new Kenya Wildlife Service 'Community Development Programme', in which local farmers and pastoralists share in nature park revenues, has recently proved to benefit both local people and the biosystem in which they live. In adjacent areas, such as the study of origins and development of indigenous agricultural knowledge systems in Kenya, results of efforts to involve local people and their knowledge in such programmes certainly look promising (Leakey and Slikkerveer 1991).

As Eugène Dubois was one of the first to substantiate the place of our species both during his journey of discovery to The Netherlands East Indies and in his ecological work on *De Bedelaer*, the understanding of human evolution in its ecological context now seems to hold an answer for our time to come.

In Louis Leakey's historic words on the occasion of a lecture in California:
"The past is the key to our future".

References

Bowler, P.J. *The Fontana History of the Environmental Sciences* (London 1992)

Campbell, B. *Human Ecology* (London 1983)

Darwin, Ch. *The Descent of Man, and Selection in Relation to Sex* (London 1871)

Darwin, Ch. *The Origin of Species by Means of Natural Selection* (London 1859)

Dubois, E. *Palaeontologische Onderzoekingen op Java*, Verslag van het Mijnwezen (1893)

Dubois, E. *Pithecanthropus erectus, eine menschenähnliche Uebergangsform aus Java* (Batavia 1894)

Dubois, E. The Place of 'Pithecanthropus' in the Genealogical Tree, *Nature* 53 (1895)

Dubois, E. Zur Morphologie des Larynx, *Anatomischer Anzeiger* 1 (1886)

Dubois, Jean M.F. *Trinil, A Biography.of Prof.Dubois, the Discoverer of P.e.* (unpublished)

Dubois, E. On the Gibbon-like Appearance of Pithecanthropus erectus, *Proceedings* 38 (1935)

Goodall, J. In the Shadow of Man (London 1971)

Goodman, M. Molecular Evidence of the Ape Subfamily Homininae, *Evol.Perspectives* (1986)

Gould, S.J. and R.W. Purcell *Finders, Keepers. Eight Collectors* (London 1991)

Haeckel, E. *Generelle Morphologie*. (Berlin 1866)

Haeckel, E. *Natürliche Schöpfungsgeschichte* (Berlin 1868)

Haeckel, E. *Systematische Phylogenie der Wirbeltiere (Vertebrata)* (Berlin 1894-1896)

Heimans, E. *Uit ons Krijtland* (Amsterdam 1911)

Huxley, T. H. *Man's Place in Nature* (London 1863 and 1893-1894)

Isaac, G. The Archaeology of Human Origins, *Advances in World Archaeology*, Vol. 3(1984)

Jacob, T. cf. G.H.R. von Koeningswald: De eerste Aapmensen in Azie, in: *Evolutie van de Mens* (Maastricht 1981)

Johanson, D.C. and T. White A Systematic Assessment of Early African Hominids, *Science* (1979)

Reader, J. *Missing Links: The Hunt for Earliest Man* (London 1988)

Koeningswald, G.H.R. von and F. Weidenreich The Relationship between Pithecanthropus and Sinanthropus, *Nature* 144 (1939)

Leakey, L.S.B. , P. V. Tobias and J. R. Napier A New Species of the Genus Homo from the Olduvai Gorge *Nature 202* (1964)

Leakey, M.D. *et al.* Fossil Hominids from the Laetoli Beds, *Nature* 262 (1976)

Leakey, R.E. *The Making of Mankind* (New York 1981)

Leakey, R.E. *Human Origins* (London 1982)

Leakey, R.E. Foreword, in: B. Theunissen *Eugène Dubois and the Ape-Man from Java* (London 1989)

Leakey, R.E. and R. Lewin *Origins* (New York 1977)

Leakey, R.E. and R. Lewin *Origins Reconsidered. In Search of What makes Us Human* (London 1992)

Leakey, R.E. and L.J. Slikkerveer *Origins and Development of Agriculture in East Afica* (Ames 1991)

Lemaire T. De Bedelaar van Prof. Dubois. Geschiedenis van een Landgoed *Rondom het Leudal* 2/3, 5-12 (1977-1978)

Lewin, R. *Bones of Contention.Controversies in the Search for Human Origins* (New York 1987)

Lewin, R. *In the Age of Mankind* (Washington 1988)

Lewin, R. *Human Evolution. An Illustrated Introduction* (Boston 1989)

Rightmire, G.P. *The Evolution of Homo Erectus* (Cambridge 1990)

Theunissen, B. *Eugène Dubois and the Ape-Man from Java* (London 1989)

Thysse, Jac P. *Waar Wij Wonen* (Zaandam 1937)

Voûte, A.M. and P.H.C. Lina De Vleermuistorens van Dr. E. Dubois, *Natuurh. Maandblad* 72 (1983)

Walther, J. and G. Heberer Im Banne Ernst Haeckels: Jena um die Jahrhundertwende (Göttingen 1953)

Zagwijn, W.H. *Aspects of Pliocene and Early Pleistocene Vegetation in The Netherlands* (1960)

Illustration Credits

The authors would like to gratefully acknowledge all those institutions and individuals for the photographs and illustrations reproduced in this book. Whilst every effort has been made to credit all photographers, publishing-companies and other sources, the publisher regrets any omission or error.

Abbreviations used:

KITLV: Royal Institute of Linguistics and Anthropology, Leiden. NFKWS: Netherlands Foundation for Kenya Wildlife Service, Leiden. NMK: National Museums of Kenya, Nairobi. NNM: National Museum of Natural History, Leiden. PCF: Pithecanthropus Centennial Foundation, Leiden. RUL: Leiden University, Leiden. RUU: Utrecht University, Utrecht.

Cover photograph:

G.H. Schubert's (1882) illustration of primates, including the chimpanzee (*S. troglodytes*) , Asian orang-utan (S. satyrus), long-arm gibbon (*S. hylobates*), and siamang (*Hylobates syndactilus*). Loek A. Zuyderduin, RUL, Courtesy of the Exhibition NNM.

Introduction.

0.1-0.2; 0.8: Loek A. Zuyderduin, RUL. Courtesy of the Exhibition NNM. 0.3: Courtesy of Dr. D.A. Hooijer, Oegstgeest. 0.4: Dr.L.J. Slikkerveer, RUL. 0.5: Ms. I. Henneke, NNM. 0.6: Courtesy of Dr. J. de Vos, NNM. 0.7: Verkade's Publications, Zaandam, Watercolour by C. Rol. 0.9: Reproduction of painting of Zdenêk Burian, NNM. 0.10: Courtesy of PCF. 0.11: Loek A. Zuyderduin, RUL, Courtesy of NFKWS. 0.12: Courtesy of the Exhibition NNM.

1. The Discovery of Pithecanthropus Erectus.

1.1; 1.3; 1.5; 1.8: Loek A. Zuyderduin, RUL. Courtesy of the Exhibition NNM. 1.2; 1.7; 1.9-1.10: Ms. I. Henneke, NNM. 1.4; 1.11-1.13: Dr. L.J. Slikkerveer, RUL. 1.6: Courtesy of Mrs. A. Hooijer, Zutphen.

2. Evolution and the Emergence of Ecology.

2.1; 2.6; 2.8: Ms. I. Henneke, NNM. 2.2-2.4; 2.7; 2.9; 2.10: Library RUL. 2.5; 2.11-2.14: Loek A. Zuyderduin, RUL. Courtesy of the Exhibition NNM. 2.15-2.18: Loek A. Zuyderduin, RUL. Courtesy of the Exhibition NNM and *Prentenkabinet* RUL. 2.19-2.20: Loek A. Zuyderduin, RUL. Courtesy of the Exhibition NNM and *Centraal Bureau voor Genealogie*, The Hague. 2.21: Loek A. Zuyderduin, RUL. Courtesy of the Exhibition NNM and Dargaud Benelux, Brussels. 2.22: Loek A. Zuyderduin, RUL. Courtesy of the Exhibition NNM and the Library of Vienna.

3. Dubois' Passion for the 'Missing Link'.

3.0: Dr. L.J. Slikkerveer, RUL. Courtesy of the Exhibition NNM. 3.1; 3.3; 3.5; 3.16-3.18: Courtesy of Mrs. A. Hooijer, Zutphen. 3.2: Courtesy of Dr. C.A.W. Korenhof, Utrecht. 3.4: Dr. L.J. Slikkerveer, RUL. Courtesy of Dr. E. F. L. Dubois, Voorburg. 3.6-3.7: Rijnja Repro, Amsterdam. Courtesy of the Exhibition NNM. 3.8-3.9: Library Dr. L.J. Slikkerveer, RUL. 3.10: Loek A. Zuyderduin, RUL. Courtesy of the Exhibition NNM. 3.11: Uitgeverij Waanders b.v., Zwolle, *Amsterdams Historisch Museum*, Amsterdam, and authors. 3.12-3.15: Library RUL. 3.19: Ms. I. Henneke, NNM.

4. *Lemuria*: Journey to The Netherlands East Indies.

4.1: Nelles Verlag GmbH, Munich 4.2; 4.4-4.5; 4.10-4.20: Loek A. Zuyderduin, RUL. Courtesy of the

Exhibition NNM. 4.3; 4.8-4.9: Library KITLV. 4.6: Courtesy of Mrs. A. Hooijer, Zutphen. 4.7: Courtesy of Dr. B. Theunissen, RUU. 4.21: Ms. I. Henneke, NNM. 4.22: Dr. L.J Slikkerveer, RUL.

5. Disclosure and Debate: Man-Ape or Ape-Man.

5.1; 5.15; 5.23: Courtesy of Mrs. A. Hooijer, Zutphen. 5.2: Courtesy of Dr. H. Siwon, Surabaya. 5.3: Dr. L.J. Slikkerveer, RUL. 5.4; 5.7: Courtesy NNM. 5.18-5.19: Ms.I. Henneke, NNM. 5.5-5.6; 5.8; 5.10-5.11; 5.16; 5.20: Loek A. Zuyderduin, RUL. Courtesy of the Exhibition NNM. 5.9: Dr. L.J. Slikkerveer, RUL. Courtesy Mrs. A. Hooijer, Zutphen. 5.12: Courtesy of Dr. L.B. Holthuys, NNM. 5.13; 5.22: Courtesy of Dr. J. de Vos, NNM. 5.14 a+b; 5.17: Courtesy of Dr. B. Theunissen, RUU. 5.21: Science Press, Beijing.

6. Africa: The 'Cradle of Mankind'.

6.1: U.S. Geological Survey, Eros Data Center. 6.2; 6.7; 6.16-6.17; 6.19-6.20: Loek A. Zuyderduin, RUL. Courtesy of the Exhibition NNM. 6.3: Reproduction of painting by Zdenêk Burian, NNM. 6.4: Based on *Scientific American* 1979. 6.5; 6.10: Collection Dr. L.J. Slikkerveer, RUL. 6.6: Kal Muller, Bareo, Singapore. 6.8-6.9: Ms. I. Henneke, NNM. 6.11-6.13; 6.22: Courtesy of NMK, drawing and design by Laura Tindimubona. 6.14: Robert I.M. Campbell, NMK. 6.15: Courtesy of the Ethiopian National Museum, Addis Ababa. 6.18: Courtesy of NMK. 6.21; 6.23-6.24: Sherma BV, London.

7. *De Bedelaer*: New Perspectives on 'Man's Place in Nature'.

7.1: Verkade's Publications, Zaandam, Watercolour by C. Rol. 7.2-7.3: Library RUL. 7.4, 7.7; 7.28; 7.31: Loek A. Zuyderduin, RUL. Courtesy of the Exhibition NNM. 7.5-7.6: Ms. I. Henneke, NNM. 7.8; 7.14: Courtesy NNM. 7.9-7.10; 7.26: Dr. L.J. Slikkerveer, RUL. 7.11; 7.29-7.30: Courtesy of Dr. John de Vos, NNM. 7.12; 7.16; 7.25; 7.27; 7.32 a+b: Courtesy of Mrs. A. Hooijer, Zutphen. 7.13: Dr. J. de Vos, NNM/Teylers Museum. Courtesy of Dr. W.H. Zagwijn. 7.15; 7.17-7.21: Loek A. Zuyderduin, RUL. Courtesy of *Rijks Herbarium*, RUL. 7.22: Drawing by E. Heimans. Courtesy of Dr. John de Vos, NNM. 7.23: B.M.F. Collet's painting of 1958. Courtesy of NNM. 7.24: Dr. Bernard Riley, Sherborne, UK.

8. The Past is the Key to Our Future.

8.1; 8.6-8.7: Drs. Marvin E. Egberts, Bilthoven. 8.2: Courtesy of Gombe National Park, Tanzania. 8.3 a+b; 8.4: Mrs. Ellen Egberts-Arnold, Bilthoven. 8.5: Collection Drs. Marvin E. Egberts, Bilthoven. 8.8-8.9: Collection Dr. L.J. Slikkerveer, RUL. 8.10: Loek A. Zuyderduin, RUL. Courtesy of the Exhibition NNM. 8.11: East African Wildlife Society, Nairobi. 8.12: P. Chichester, National Audubon Society, and John Wiley & Sons Inc., New York. 8.13: N. Tabu, Courtesy of NMK. 8.14: Courtesy of Dr. Pierre-Fr. Puech, Nimes, France.

INDEX

References in **bold** type indicate illustrations, *i.e.* the page numbers are those of the captions, not of the illustrations themselves.

A

Aardrijkskundig Genootschap 101
Acheulian 118; 132
Acidification of mist and rainwater 143
Alberdingk Thijm, J.A. 67
Anatomy 36; 48; 67; 68; 69; 70; 103; 139
Ancestry 36; 48; 76; 103; 107; 110; 113; 118;165
Anderson, J.G. 81; 106; 138
Andrews, R.C. 81, 103
Anthropopithecus 91; 93; 99; **94**
Anthropos 40
Ape-Man 11; 13; 15; 20; 29; 46; 91; 93; 99; 100; 102; 103; 110; 135; 163; **11; 13; 21; 94**
Arambourg, C. 120
Aristotle 35
Arthritis 48
Australopithecus 17; 76; 116; 118; 120; 121, **122; 126**

B

Bat-roosts 160; **156**
Bat-towers 160; 162; **157**
Beagle 35
De Bedelaer 5; 17; 18; 20; 137; 139; 146; 155; 159; 160; 162; 163; 178; 191; 194; **142; 144; 145; 147; 148; 162; 163**
Behavioural 42; 167
Belgisch-Nederlandsche Vereeniging 162; **157**
Bernard Ngeneo 121; **123; 124**
Biodiversity 5; 170; 176
Bipedality 50
Bishop Ussher 36
Black, D. 106; 125
Bleeker, P 76
Blumenbach, J.F. 42
Boule, M. 113
Bowler P.J. 45; 137
Brain cast 17
British Association for the Advancement of Science138
Broken Hill Man 135
Broom, R. 116; 118
Brouwers, P. 67
Buckland W. 46; **40**
Buffon, G.L.L. 42; 54
Bushmen 37

C

Calwer, C. G. 62
Campbell, C.H.R. 160, **156**
Candelabra model 135; **134**
Centennial 18; 20; 93; **18; 20**
Cephalization 13; 15; 17; 50; 62; 103; 107; 160; **105**
Chou Kou Tien 106; **109**
Classifications 37; 42
Collectie Dubois 17; 67; 93; 99; 139; **140**
Commissie tot Bevordering van Natuurkundig Onderzoek 77
Coniferae and deciduous trees 159
Conservation 5; 17; 198 162; 163; 177
Convention of International Trade in Endangered Species (CITES) 178
Copernicus, 35
Count Samuel Teleki 121
Cradle of Mankind 5; 17; 73; 113; 130; 165
Creation 36; 67; 155; 159
Cro-Magnon 48; 113
Cultural diversity 176
Culture 37; 48; 50; 176
Cunningham, D.J.15; 101
Cuvier, G. 42
CuypersP.J.H.,67

D

Dart, R.A. 116
Darwin, Charles.5; 11; 13; 15; 35; 36; 42; 45; 48; 73; 91; 99; 130; 138; 165; **34; 36; 46; 138; 139**
Departement van Onderwijs 77; 81
Dragon Bone Hill 106
dragon bones 81; 106
Dupont 48
Dutch Golden Age 62

E

Ecological 13; 17; 20; 42; 62; 111; 113; 135; 167; 170; 178; **39; 40; 163**
Ecology 5; 13; 15; 17; 18; 20; 35; 48; 111; 137; 138; 139; 167; **16**
Ecosystem 46; 170; 177
Embryology 36; 45; 48; 50; 137
Encephalization 50
Environment 17; 18; 35; 45; 86; 120; 137; 138; 143; 147; 160; 165; 176; 170
Ethnographic analogy 177
Ethnology 50
Evolutionary theory, 35; 36; 48; 103; **39**
Eysden 13; 53; 54; 69; **53; 54; 55**

F

Family-tree 15; 46; 48; 93; 99; 110; 120; 165; **46; 48; 49; 57; 97; 106; 127; 128; 131**
Fauna(l) 13; 17; 35; 73; 76; 77;

79; 86; 87; 88; 99; 139; 143; 162; 163
Faunas from Java **87**
Fen(s) 27; 143; 146; 147; 160; 162; **136; 142; 148**
Fertilization of water 155
First Family 125
Flower, W.H. 102
Footprints 125; **125**
Forestry 159; 162
Fort Van den Bosch, 27; **25**
Fossey, D. 167
Fourth International Congress of Zoology 103; **105**
Fuhlrott, J.C. 48
Fürbringer, M. 68; 69; 70

G
Galapagos Islands. 35
Gap 37; 69; 99; 167
Gegenbauer, C. 68
Genealogical 46; 48; 50; **56**
Gibbon 15; 76; 87; 99; 100; 102; 107; **96**
Goodall, J. 166
Goodall Institute **167**
Goodman, M.130
Gould, S.J. 67
Granger, W. 81
Groeneveldt, W.P. 76; 77
Grote Postweg (Great Mailroad) 27

H
Habitat.5; 54; 162
Haeckel 11; 13; 15; 45; 46; 48; 67-69; 73; 93; 99; 100; 103; 130; 137; 138; **40; 42-45**
Handy Man 118
Heberer, G. 137
Heidelberg Man 106; 135
Hoff, J.H. van 't68
Holistic 13; 19; 45; 137; 159
Homo caudatus 42
Homo erectus 13;17; 18; 76; 86; 87; 107; 121; 125; 130; 132; 135; 166; **19; 124; 129; 132-134**
Homo habilis 17; 76; 118; 120-121; 129; 166; **123**
Homo heidelbergensis 113
Homo primigenius 100
Homo sapiens 18; 37; 42; 99; 129; 132; 135; 160; 165; 170; 176
Homo soloensis 106; 110
Homo troglodytes 42
Homo wajakensis 80
HookerJ. 36
Hottentots 37
Howell, F.C. 120
Howells,W.W. 135
Human milieu 18
Huxley, Thomas H. 5; 11; 13; 18; 36; 37; 45; 48; 67; 69; 130; **37**

I
Indigenous 162; 177; 178
Indisch Comité van Wetenschappelijk Onderzoek 77
Indonesian Geological Survey 113
Inevitability 167
International Botanical Congress 18; 138
Isaac, G. 166
Ivory (ban) 178; **174; 175**

J
Jacob, T. 113
Jaekel, O. 15
Jentink, F.A. 77
Johanson, D.C. 125; **126; 127**
Julius, W.H. 67
Junghuhn, F. 81, **82**

K
Kallilimne 146; 155
Kate, H.F.C. ten 102
Keith, Sir A. 15; 101; **100**
Kenya Wildlife Service 20; 178; **174-175**
Kimeu, Kamoya 121; **129; 133**
Kinship 48; **46**
Koeningswald, G.H.R. von 107; 113; 137
Königliche Bayerische Academie 103
Koninklijk Nederlandsch-Indisch Leger 23; 76; 99
Koninklijke Natuurkundige Vereeniging 79
Koninklijke Nederlandsche Academie 15; 111; 143
Koobi Fora 120; 121; 125; **112; 123-124**
Kramps, J.M.A. 67
Krause, W.15; 100; 102
Kriele, G. 23; 28; 29; 81; 86; **25**
Kroesen, R.C. 76

L
La Chapelle-aux-Saints 113, **115**
Laetoli 125; **125**
Lake Turkana 120; 121; 166; **112; 122-124; 129; 133**
Lamarck, J.B. 35; 45
Large-scale plantation 155
Larynx 69; 166
Leakey, Louis S.B. 20; 118; 121; 165-166; 178; **117**
Leakey, Mary 118; 121; 125; 129; 165; **118; 125**
Leakey, Richard E. 107; 120; 121; 125; 129; 130; 165; 176; **20; 122; 129; 131; 175**
Leiden University 20; 67
Lemaire, T. 18; 143; 147; 160; 163
Lemuria 5; 13; 69; 73; 93; 99; 130; 137
Lewin, R. 36; 125; 130; 165; 176
Lida Adjer Cave 76
Limburg 15; 17; 20; 53; 54; 62; 69; 111; 139; 143; 146; 147; 155; 160; 162; 163; **16; 55; 136; 147; 154; 155**
Lina, P.H.C. 162
Linnaeus 42; **40**
Lucy 125; **126**
Lydekker, r. 15; 73; 91; 93; 99; 101
Lyell, C. 36

M
Maas River 13; 53; 54; 62; 143; 155; 55
Malay Archipelago 76; **96**
Malthus, T. 35
Man the Hunter 166
Management of water 155
ManouvrierL. 15, 100; 102; **104**
Marsh, O.C. 15; 101; 102; **45**
Martin, K. 15; 77; 101
Mergel 54
Microcosm 63
Migration 87; 132; 135; 166
Milne Edwards, A. 102
Minerals 155
Missing Link 5; 11; 13; 18; 20; 80; 91; 93; 100; 110; 125; 129-130; 137; 166; **13; 94; 97**
Mitochondrial DNA 135; 166
Modjokerto 107
Molar 11; 87; 91; 99; 100; 102; 106; 118; **95; 98-99**
Molecular 17; 130; 135; 165
Molengraaff, G.A.F. 143
Monism 45; 137
Monogenist 50

N
Napier J. 118

National Geographic Society 120
National Museum of Natural History 11; 17; 20; 67; 77; 99; 139; **140-141**
National Museums of Kenya 20
Natural Selection 35; 36
Natuurmonument 17; 162
Natuurpark 155; 162
Neanderthal 48; 69; 76; 100; 101; 102; 113; 116; 129; 166; **114-115**
Nederlandsch-Indische Geologische Dienst 113
Nederlandsche Koloniale Landbouw school 146
Nederlandsche Vereeniging tot Behoud van Natuur-monumenten 162
Netherlands East Indies 13; 20; 23; 67; 68; 70; 73; 76; 77; 93; 102; 130; 165; 166; 178; **70-71; 90**
Netherlands Hydrobiological Association 147
Newton, E.T. 35
Ngawi 27; 28; **25**
Noah's Ark' 135; 166; **134**

O
Oecologie 20; 137
Olduvai 118; 120; 121; 125
Omo Valley 120
Oppenoorth, W.F.F. 106
Owen, R. 37

P
Padang 76
Pati Ajam Mountain 81
Pei Wenzhong 106
Peking Man 81; 106; 107; 116; 135
Pesthuys 20; **11; 21; 141**
Petit, August 100
Phylogenetic 13; 69; 99; 100; 103; 137; **97; 106**
Physiology 68; 137; 138
Piltdown Man 113
Pithecanthropus 13; 17-18; 46; 69; 93; 99; 101; 106-107; 110; 113; 130
Pithecanthropus alalus 46
Pithecanthropus erectus 5; 11; 23; 103; 139; **12-13; 15; 30-31; 33**
  change of name 107
  disclosure of.29
  emphasis on 163
  gibbon-like appearance of 15
  Haeckel's publication 100
  monument.91
  publication 99
Pithecos 46
Place, Thomas 68
Polygenetic evolution 50
Primate studies 101; 166
Prinses Amalia 70; 76; **70-71**

R
Raden Saleh 81
Rariteiten Cabinet 67; **62-63**
ReaderJ. 68; 110
Rietschoten, B.D. van 79; 80; **78**
Rijks Hoogere Burger School 67
Rijks Kunstnijverheidsschool 68
Rijksmuseum voor Natuurlijke Historie 77
Rijksnormaalschool voor Tekenonderwijs 68
Rosen, J. 67
Ryckholt 69

S
Sangiran 107; 113
Schlosser, M. 106
Schoetensack, O. 113
School of Morphology 48; 68
Schroevers, P. 147
Schwalbe, G.A. 100
Selenka, M. 103; **108**
Sinanthropus 106
Sint Pietersberg 62
Siwalik 73; 86; 88; 93; 99; **98-99**
Sixth Extinction 170; 176
Skullcap 48; 91; 99; 100; 101; 102; **30**
Slikkerveer, L.Jan 178; **91; 140**
Slikkerveer, Mady **20**
Sluiter, C.P.79
Société d'Anthropologie 100
Solo Man 106
Solo River 11; 23; 27; 86; 88; 91; 93; **24-26; 29; 32; 88; 92; 102**
Sondaar, P.Y. 87
Steijn, L. 143
Sularso **91**
Suringar, W.F.R. 62
Swaving, C.76

T
Taung Child 116
Tegelen 17; 143; **144-145; 155**
Terrestriality 50
Teylers Museum 67
Theunissen, B. 15; 48; 70; 103; 111; 137; **81**
Thigh-bone 11; 93; 99; 100; **31; 95**
Third International Congress of Zoology 102; **102**
Thysse, Jac.P. 17; 162-163; **136**
Tienhoven, van 163
Timmerman, J.A.C.A. 101
Tobias, P.V. 118; **91**
Toeloeng Agoeng 80; **80**
Treub, M. 77
Trinil 20; 76; 87-88; 91; 93; 101; 103; 110; **22; 24-26; 28; 30-33; 83-85; 88-89; 91; 108**
Turkana Boy 76; 129; 135, **129**
Turner, W. 15; 101

U
Universitas Padjadjaran 20
University of Amsterdam 13; 67; 68; 69; 111; 139; **111; 160**
University of Munich 106
Untouched nature 160
Utrecht University 68

V
Verbeek, R.D.M. 79
Vereenigde Oost-Indische Compagnie (VOC) 63
Victorian 11; 35; 67
Virchow, R. 15; 48; 100; 102; 103
Vos, J. de 67; 87; **140**
Voûte, A.M. 162
Vries, H. de 68
Vuyck, L. 146; 162

W
Waals, J.D. 68
Wadjak 79; 80; 81; 87; 106; **78**
Walker, Alan 121
Wallace, A.R. 13; 35; 36; 69
Walther, J. 137
Weidenreich, F. 106; 107; **108-110**
Westhoff, V. 147; 160
White, T. 125
Wildlife 135; 177-178
Winter, A. de 23; 81; 83; **25**
World Exhibition in Paris 103; **107**

Z
Zagwijn, W.H. 143;
Zinjanthropus 118; **118**